Paleozoic Fossil PLANTS

Bruce L. Stinchcomb

4880 Lower Valley Road • Atglen, PA 19310

Other Schiffer Books by the Author:
Paleozoic Fossils, ISBN: 978-0-7643-2917-3, $29.95
More Paleozoic Fossils, ISBN: 978-0-7643-4030-7, $29.99

Library of Congress Control Number: 2013942830

Designed by RoS

Type set in DIN/Arrus BT

ISBN: 978-0-7643-4327-8
Printed in The United States of America

Published by Schiffer Publishing, Ltd.
4880 Lower Valley Road
Atglen, PA 19310
Phone: (610) 593-1777; Fax: (610) 593-2002
E-mail: Info@schifferbooks.com

Taxonomic Disclaimer

This work covers a wide variety of fossils, often identifying them to genus and species (precise taxonomic position). In some of his previous works on fossils, the author has been criticized because the taxonomy presented was not "up to date." Modern paleontology, especially taxonomic paleontology, has become very specialized and focused. Workers dealing with a specific taxa often have trouble keeping abreast with current literature dealing with the group in which they specialize. The world of life in the geologic past and preserved as fossils, like that of life in the present, was rich and diverse. To expect taxonomic currency in general works such as this, works covering a broad range and scope of fossils, is to expect the impossible. Paleontologic taxonomy is a "work in progress," names given in a particular paleontologic source should be looked upon as "handles" by which a person interested and concerned with a particular fossil can gain taxonomic access. This is especially the case with use of the Internet as a fossil name can be used as a search word and its current validity often determined as well as current information regarding it if the name has been placed in synonymy.

Acknowledgments

As with the author's other books on fossils, numerous persons have contributed in various ways toward the execution and completion of this work. Specifically, the author would like to thank the following serious collectors for their interest, appreciation, and specimens of fossil plants: Nick Angeli, Chris Braught, Patricia Eicks, Matt Forir, Gerald Gunderson, Steve Holley, Tim Northcutt, Steve Pavelsky, John Stade, Barry Sutton, Richard Thoma, and Dennis Whitney.

The author would also like to acknowledge the following persons in academia for various encouragements and assistance, which directly, indirectly, or inadvertently led to this work: Henry Andrews and Dorothy J. Echols—both formerly of Washington University in St. Louis, the late Alfred Spreng of Missouri S&T (Science and Technology and formerly University of Missouri-Rolla), the late Ellis Yochelson of the U.S. National Museum and USGS, and James J. Jennings of University of Illinois, Carbondale.

He would also like to acknowledge for various assists: Martin Connelly, Thomas Foeller, Scott George, Pete Kellams, Jeff Snyder, and Mike Wiskoski.

Lastly, he would like to acknowledge the effort and encouragement put toward his various interests years ago by his parents, Leonard and Virginia Stinchcomb, efforts that included both helping with the acquisition of relevant literature and the collecting of Paleozoic fossil plants.

Contents

Chapter One
Paleozoic Fossil Plants
In The Beginning!

Geologic Time Scales

In The Beginning—or Eozoon Canadense

Beginnings are always exciting! After fossils were realized to be what they really are, evidence of life that existed in the earth's geologic past, a major quest began for discovery of the earliest records of life in this geologic past. One of the results of this quest was the finding of Eozoon Canadense, "The Dawn Animal of Canada." Eozoon was believed to be a giant rhizopod or protist. It was found in rock layers far below (and hence older) than any strata yielding previously known fossils. Eozoon came from rocks of Precambrian age of the Canadian Shield of Quebec. This hypothesis started a controversy as to its biogenicity, a controversy that continued until the beginning of the twentieth century. At that point, a consensus of scientific opinion considered Eozoon as a pseudofossil.

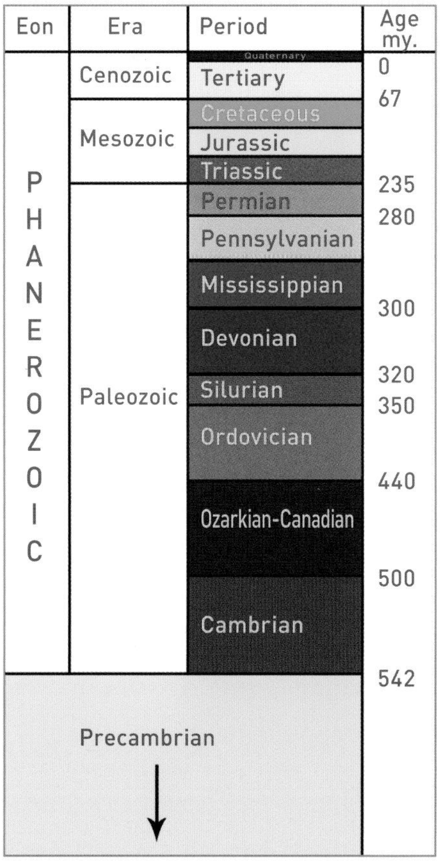

Eon	Era	Period	Age my.
PHANEROZOIC	Cenozoic	Quaternary	
		Tertiary	0
	Mesozoic	Cretaceous	67
		Jurassic	
		Triassic	235
	Paleozoic	Permian	280
		Pennsylvanian	
		Mississippian	300
		Devonian	320
		Silurian	350
		Ordovician	440
		Ozarkian-Canadian	500
		Cambrian	542
Precambrian ↓			

Geologic time scale-I: the Paleozoic Era includes Cambrian through Permian periods. The Ordovician includes the Ozarkian Period, as this has been one of the author's subjects of interest.

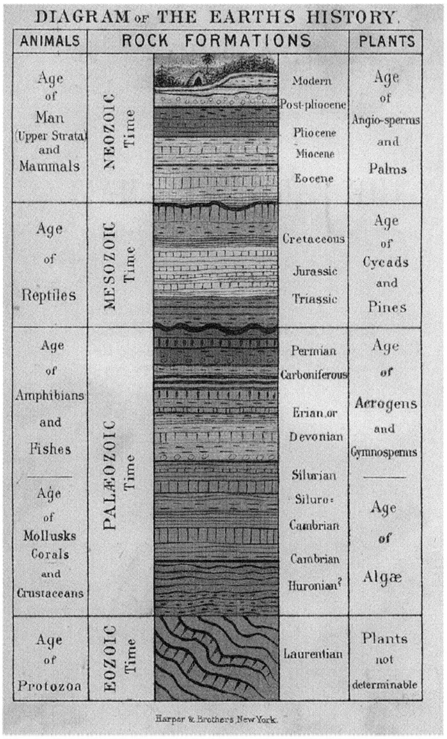

DIAGRAM OF THE EARTHS HISTORY.

ANIMALS	ROCK FORMATIONS		PLANTS
Age of Man (Upper Strata) and Mammals	NEOZOIC Time	Modern, Post-pliocene, Pliocene, Miocene, Eocene	Age of Angio-sperms and Palms
Age of Reptiles	MESOZOIC Time	Cretaceous, Jurassic, Triassic	Age of Cycads and Pines
Age of Amphibians and Fishes	PALÆOZOIC Time	Permian, Carboniferous, Erian or Devonian	Age of Acrogens and Gymnosperms
Age of Mollusks Corals and Crustaceans		Silurian, Siluro-Cambrian, Cambrian, Huronian?	Age of Algæ
Age of Protozoa	EOZOIC Time	Laurentian	Plants not determinable

Harper & Brothers New York.

Geologic Time Scale-II: the geologic time scale evolved and was compiled primarily during the nineteenth century. This is an example from J. William Dawson's *The Story of Earth and Man*, a late nineteenth century discourse of earth history. Note that the Ordovician is missing, being represented by Siluro-Cambrian. Also note the Huronian, thought at the time by some geologists to be a geologic period before the Cambrian. It was hypothesized as having Paleozoic-like strata in which it was thought would be found fossils similar to, but preceding, those of the Cambrian. The strata of the Huronian, like that of the Cambrian of North America, was relatively unmetamorphosed. Some mid-nineteenth century geologists even claimed to had found trilobites in Huronian strata. In the mid-twentieth century, strata like this would yield the Ediacarian fossils, which appear to be weird and unrelated (according to some paleontologists) to the life represented by Cambrian fossils. Note on the "scale" the dominant animals on the **left** (including "Age of Protozoa" for Eozoic time) and plants on the **right**.

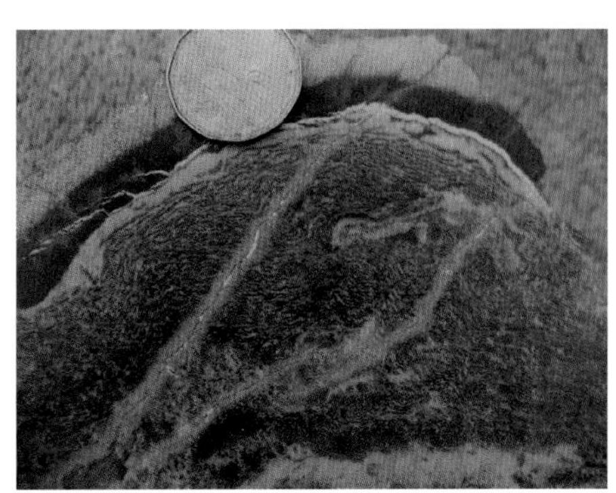

Polished section of Eozoon canadense, Grenville Marble, Quebec.

Naturally weathered surface of ***Eozoon canadense***. This is the "Protozoa" in Dawson's Geologic Time Scale. It was considered by Dawson, as well as many other geologists of the time, as the earliest evidence of life and the **oldest fossil**. Eozoon was thought (by those convinced it was organic) to be a gigantic rhizopod or protozoan. Eozoon is now recognized to be a pseudofossil, a product of intense metamorphism of ancient Precambrian limestone. Its nearest claim to biogenicity is that it may represent severely metamorphosed stromatolites present in the limestone, which were metamorphosed to form the Grenville Marbles in which Eozoon was found.

Fusilinid: a giant fossil rhizopod or protozoan (protista). Fusilinids are protozoans possessing a calcareous "shell" referred to as a "test." Previously they were considered to be one-celled **animals**. They are now placed in the kingdom protista (single celled eukaryotes) and, hence, not considered animals anymore.

In addition to this inferential evidence, however, one well-marked animal fossil has at length been found in the Laurentian of Canada, *Eozoon Canadense*, (fig. 7), a gigantic representative of one of the lowest forms of animal life, which the writer had the honour of naming and describing in 1865—its name of "Dawn-animal" having reference to its great antiquity and possible connection with the dawn of life on our planet. In the modern seas, among the multitude of low forms of life with which they swarm, occur some in which the animal matter is a mere jelly, almost without distinct parts or organs, yet unquestionably endowed with life of an animal character. Some of these creatures, the Foraminifera, have the power of secreting at the surface of their bodies a calcareous shell, often divided into numerous chambers, communicating with each other, and with the

Discussion of Eozoon Canadense by J. W. Dawson, the mid- and late nineteenth century paleontologist who discovered it. After Charles Darwin published his "Origin of Species" in 1859, considerable resources were put into trying to find the ancestors of life forms represented by Cambrian fossils. It was generally thought that it would just be a matter of a "lucky find" to discover fossils that were ancestors of those found in Cambrian rocks.

THE EOZOIC AGES. 25

the sea, and covering its gelatinous body with a thin crust of carbonate of lime or limestone, adding to this, as it grew in size, crust after crust, attached to each other by numerous partitions, and perforated with pores for the emission of gelatinous filaments. This continued growth of gelatinous animal matter and carbonate of lime went on from age to age, accumulating great beds of limestone, in some of which the entire form and most minute structures of the creature are preserved, while in other cases the organisms have been broken up; and the limestones are a mere congeries of their fragments. It is a remarkable instance of the permanence of fossils, that in these ancient organisms the minutest pores through which the semi-fluid matter of these humble animals passed, have been preserved in the most delicate perfection. The existence of such creatures supposes that of other organisms, probably microscopic plants, on which they could feed. No traces of these have been observed, though the great quantity of carbon in the beds probably implies the existence of larger seaweeds. No other form of animal has yet been distinctly recognized in the Laurentian limestones, but there are fragments of calcareous matter which may have belonged to organisms distinct from Eozoon. Of life on the Laurentian land we know nothing, unless the great beds of iron ore already referred to may be taken as a proof of land vegetation.*

Dawson's description of Eozoon, in which he considers it to have formed from a gelatinous protoplasmic mass. Also being investigated at the time was Bathybius and Nummulites, the later a fossil protozoan—these were linked to Eozoon (See Steven J. Gould's "Bathybius and Eozoon" in *The Panda's Thumb*). It is interesting to note that in the outcrop of Eozoon visited by the author, Eozoon masses do occur as globs suggestive of Paleozoic fossil corals or stromatoporoids, both of which form reefs, an association no doubt noted by Dawson. It is also of note that many very ancient limestone and marble beds contain a lot of graphite (Dawson's Great Quantity of Carbon). The "Great Beds of Iron Ore," which Dawson also mentions, are BIF (Banded Iron Formation), a geochemical sediment now considered to be a biochemical record of the early evolutionary stages of the earth's atmosphere. BIF formed from the oxidization of ferrous iron dissolved in sea water and preceded the accumulation of free oxygen in the atmosphere.

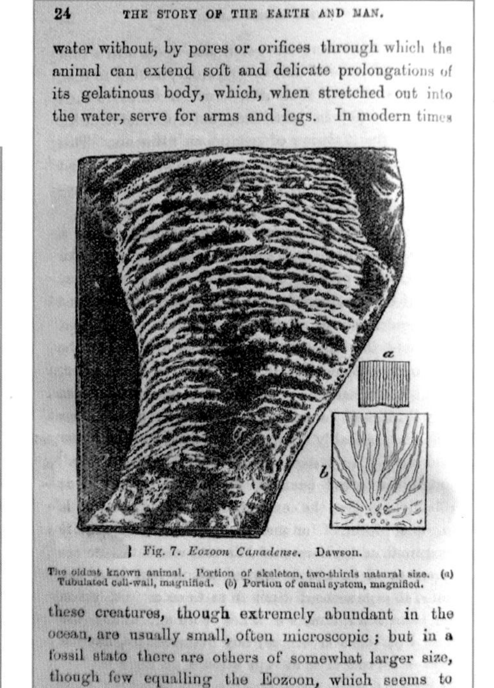

24 THE STORY OF THE EARTH AND MAN.

water without, by pores or orifices through which the animal can extend soft and delicate prolongations of its gelatinous body, which, when stretched out into the water, serve for arms and legs. In modern times

Fig. 7. *Eozoon Canadense.* Dawson.

The oldest known animal. Portion of skeleton, two-thirds natural size. (a) Tubulated cell-wall, magnified. (b) Portion of canal system, magnified.

these creatures, though extremely abundant in the ocean, are usually small, often microscopic; but in a fossil state there are others of somewhat larger size, though few equalling the Eozoon, which seems to have been a sessile creature, resting on the bottom of

Illustration of Eozoon Canadense from Dawson's *The Story of Earth and Man.*

The Oldest Fossil Plants

Fossil plants in this work include those relevant to the question, "What was the earth's earliest land vegetation like and when did it appear?" Considerable amounts of hypothesizing dwelt on this issue, although **the most direct source of information on life of the geologic past is, of course, the fossil record**. The time of appearance and type of the first land plants may always be an enigma, but definitive information on it does exist. Still, "land plant firsts" are quite "muddy" and subject to quite a lot of interpretation. Some "land plant pioneers" are serendipitous finds that occur frequently enough as carbonaceous compressions, making such "**firsts**" interesting collectibles in themselves. What carbonaceous compression found in the Early Paleozoic (Silurian, Ordovician, and even Cambrian) actually represents the earliest **land vegetation** is difficult to determine and remains an enigma as many such ancient carbonaceous compressions are from marine algae.

Paleozoic Plants

Fossil plants are a category of collectibles that can be attractive (pleasing eye candy) as well as scientifically and educationally significant. They can also be abundant enough so that there usually are sufficient specimens available for both science and collectors. As with modern plants, Paleozoic plants include both marine and freshwater forms as well as the more commonly considered land plants. Of the former, marine plants, such as various forms of algae, can be relatively common and conspicuous fossils, especially in marine strata like limestone. Fossil plants can also be locally common fossils, a fact that might be expected as plants are always major elements in communities of organisms—major elements due to the fact that they are the primary producers upon which animals (as well as representatives of other kingdoms, such as fungi) ultimately depend.

Fossil plants illustrated and discussed here are those from the Paleozoic Era, spanning hundreds of millions of years into the geologic past—a geologic past that was a world quite different from the one we know today. Fossil plants of the early part of the Paleozoic (the Cambrian, Ordovician, and marginally the Silurian period) appear to be entirely marine (so far as is known). These early Paleozoic plants consist of various types of algae, as during this time land plants (presumably) had yet to evolve. The later part of the Paleozoic Era, on the other hand, finds a plethora of land plants. This was the time of the **Carboniferous Period**, a time of land plant biomass sufficiently great to be responsible for a major portion of the earth's coal beds. Carboniferous coal (especially Late Carboniferous), to a major degree, is the coal that has played such an important part in modern industrial civilization and technology, producing a cheap and readily available energy source.

Compression fossils (where plant material is preserved as a thin film of carbon) collected from Cambrian age strata. If these fossils were found in younger rocks, they would be considered as land plants ... undeterminable ones, but land plants never-the-less. Because these occur in such ancient rocks and are not identifiable as to any specific plant type, they are not scientifically significant. This is the case with other similar occurrences like this in early Paleozoic rocks.

The Author's Early Interest in Fossil Plants

The author, as a child, became familiar with Paleozoic fossil plants and strata preserving them when outcrops of fossil plant bearing shale were discovered by him in what, at the time, was the northern edge of the St. Louis metropolitan area. This early introduction fixed his interest in these early "medals of creation." Although his main focus later became Cambrian and Precambrian fossils, Paleozoic plants have always been of interest as they record the beginnings of life on the terrestrial, or land, portion of the planet, an event that happened long after life (both plant and animal) first appeared in the sea. Already familiar with invertebrate animal fossils of similar age (Pennsylvanian or Upper Carboniferous), the finding of a horizon of nice, well preserved Late Paleozoic plants was quite exciting. All of these fossil occurrences were within bicycling distance, but the plants were especially exciting as there already was an inkling that they represented vegetation that existed near the beginning of land life on the planet and **beginnings are always exciting**. Considering that Upper Carboniferous (Pennsylvanian) rocks cover a large portion of the U.S. Midwest, the significance of such a find diminished with later knowledge that fossil plants similar to these occur sporadically over a large area; but, thinking of the thrill of finding such a treasure to this day is still exciting. Pennsylvanian age strata (the author prefers this to the Late Carboniferous of European usage) locally can contain shale layers bearing fossil plants in many places in the Midwest U.S. Usually these plants are those that were part of the "coal age" flora, the earth's first endowment of really abundant and dense vegetation. Rarer are those plant fossils of the same age representing drier, more upland floras.

Paleozoic fossil plants from worldwide localities are presented here; however, as is the case with most works dealing with fossils, there is a certain bias toward specimens from specific regions. In *this* work that bias exists for fossil plants of the U.S. Midwest. As has been previously mentioned, the author first became involved in fossil plants by the "hands on" collecting method. Considering the relative ease, using digital techniques, by which quality images can be reproduced, it is hoped that other serious workers in fossils, as well as others involved with geo-collectibles (be they amateurs or professionals), will showcase material from other regions of the country in the manner done here. In other words, present paleontology (or paleobotany) in a non-technical but accurate and readable way, showing that science can be fun, exciting, and aesthetic.

Fossil land plants in marine limestone containing fossil brachiopod shells. Fossils of marine animals are more common than plants as animals commonly have hard parts like shells, bones, and teeth. These structures are readily preserved as fossils. Plants, especially land plants, unless preserved as a carbon film or compression, like this, are not as readily preserved.

Fossil Marine Plants

Fossil plants in sedimentary rocks can be found with some frequency, especially in non-marine strata like river and swamp deposits. Marine plants, such as various types of marine algae, can also leave a variety of fossils, some of which have been very puzzling to paleontology. The fact that fossil plants are not as commonly seen or reported from rock strata as are animals may seem counter intuitive as plants in any ecosystem generally outnumber animals in biomass. The reason for this is that many animals, especially marine examples, produced shells or other hard parts that can be preserved in large numbers in marine strata like limestone. Plants, one the other hand, usually are made up of material that is totally organic—compounds of carbon, hydrogen and oxygen—and it takes somewhat more specific conditions to preserve this material over geologic time. The usual method for preservation being where plant material is preserved as a film of coaly carbonaceous material (compressions) or as a film of carbon or graphite. Ecologically, plants are the primary producers of today and as all of nature has to obey the laws of physics and chemistry, they also had to be primary producers in the geologic past. Their less frequent occurrence as fossils is a consequence of the vicissitudes of fossil preservation.

Plant fossils of an indeterminate type: these Late Devonian compression fossils are from plants. It is unclear, however, as to what type of plants they were, land or marine. They were considered by the author to be psilophytes (primitive leafless land plants); however, this was somewhat of a scientific guess and they may well be a type of marine algae (a sea weed). Their preservation as carbonaceous compressions suggests that they may originally have been composed of cellulose, a material usually associated with land plants. Grassy Creek Formation, Upper Devonian, Hillsboro, Missouri.

Major Stages of the Earth and Fossil Plants

Recently (2010), earth history has been placed in a novel context of major stages in the planet's evolution and development over some 4+ billion years of geologic time. Unlike other terrestrial planets of the Solar System, the Earth has undergone a series of **distinct evolutionary stages,** which have left a record and stand out over its 4.5 billion year history. These evolutionary changes were brought on as a consequence of that unique phenomena documented by the planet's paleontological record, **life**. After formation of Earth from the solar nebula, stage one consists of the "**Black-Earth-Stage,**" a time when massive meteorite impacts influenced a magma ocean composed (when it cooled) of the black mafic rock known as basalt. It was during this time that life either evolved or was introduced to the earth. This is followed by the **Red-Earth-Stage**, a period not only of continent formation but also of oxygen production atmospheric accumulation through photosynthesis. Photosynthetically produced elemental oxygen oxidized large amounts of soluble ferrous iron dissolved in the oceans of that time. This iron was oxidized to red, ferric oxide, oxidized beds of iron oxide known today as banded iron formation (BIF). These iron-rich beds today are the source of most of the iron used in the steel industry. This was followed by the "**White-Earth-Stage,**" a period of extensive glaciation. The Silurian Period of the Paleozoic Era ushered in the "**Green-Earth-Stage,**" a span of time continuing to the present where land areas of the planet were covered by land plants, the first of which were Paleozoic land plants, the subject of this book.

Some of the Earth's Oldest Rocks, Greenstone Belts: much of the rock of the early earth is associated with what are known as "greenstone belts." This ancient rock is generally black or green in color and represents some of the first stages in the planet's geologic evolution. Because black rock known as basalt predominates in the early rock record, this first stage of earthly evolution is referred to as the "**Black-Earth-Stage.**" The black rock shown here is basalt. It represents rock that originated from the earth's mantle some 3.5 billion years ago in the Archean Era. The lighter rock is granite, intruded into the black basalt or mafic rock during early activity that produced the continents. This granite contributed to part of an Archean protocontinent now found in central Manitoba, Canada.

Xenoliths (strange-rocks) in 3.2 billion year old granite. The dark masses are rocks "plucked" from walls of the chamber through which an ascending mass of molten rock traveled. Xenoliths found in very ancient terrains like this may represent some of the earth's original crust—crust that cooled just after **an intense siege of meteorite bombardment, which previously kept the earth's surface in a molten condition.**

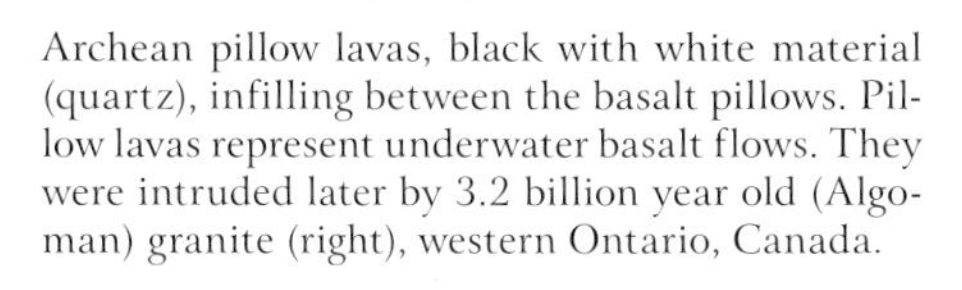

Archean pillow lavas, black with white material (quartz), infilling between the basalt pillows. Pillow lavas represent underwater basalt flows. They were intruded later by 3.2 billion year old (Algoman) granite (right), western Ontario, Canada.

Fossil threads (parallel filaments) of cyanobacteria that have been preserved by leaving a carbon film (compression fossil or organic distillation).

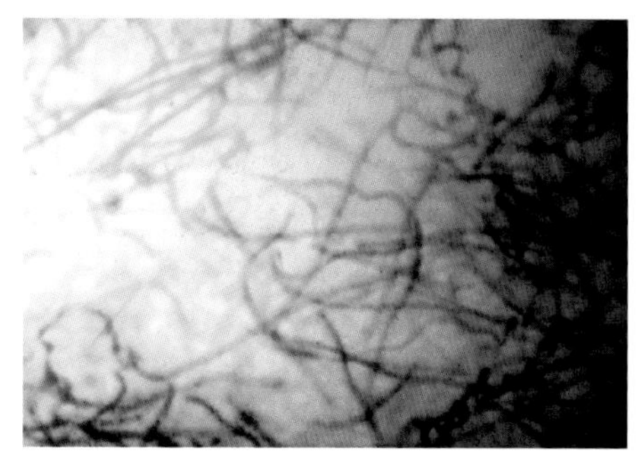

Strands (or filaments) of cyanobacteria seen under a microscope. These represent one of the more primitive and ancient forms of photosynthetic life.

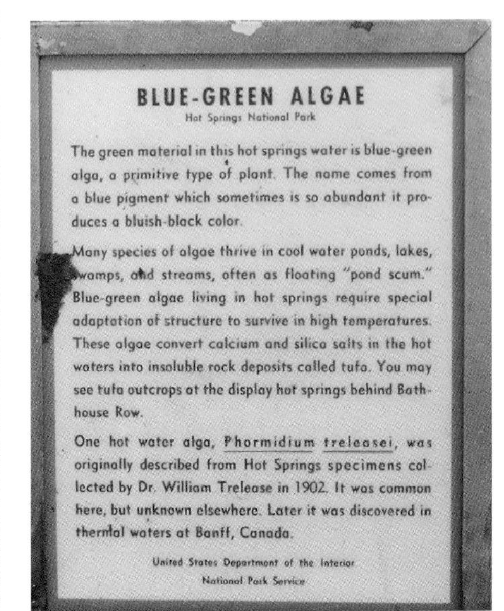

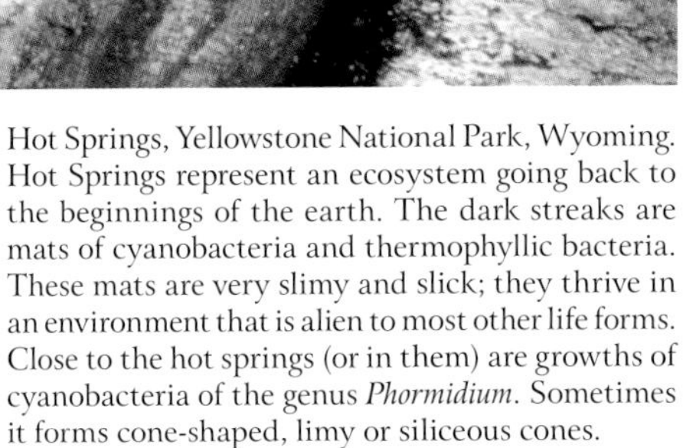

Hot Springs, Yellowstone National Park, Wyoming. Hot Springs represent an ecosystem going back to the beginnings of the earth. The dark streaks are mats of cyanobacteria and thermophyllic bacteria. These mats are very slimy and slick; they thrive in an environment that is alien to most other life forms. Close to the hot springs (or in them) are growths of cyanobacteria of the genus *Phormidium*. Sometimes it forms cone-shaped, limy or siliceous cones.

Blue-green algae, Hot Springs National Park, Arkansas: now referred to as cyanobacteria, **blue-green algae no longer is considered as a plant**, rather it is a type of **moneran**. Monerans represent a major category of life in which cells **lack a cell nucleus**. The mentioned *Phormidium* (found in the algal mats of Hot Springs) is a genus of cyanobacteria commonly associated with hot springs. The greenish material on the left side of the sign itself is blue-green algae.

Stromatolites: this is a colony of stromatolites, structures formed from photosynthetic activity of monerans, especially cyanobacteria or blue-green algae. Stromatolites like this represent the vast majority of fossils found in rock strata formed from 3.5 billion years ago to 800 million years ago. The peak of stromatolite development and abundance was during the "Red-Earth-Stage" of earth's planetary evolution.

Conophyton sp. This is a type of fossil known as a **stromatolite**. It is a structure produced by the physiological activity of primitive life forms. These particular stromatolites form as a series of nested cones, hence the name Conophyton = cone-like "plant." *Conophyton* often forms in hot springs like that of the previous photo. This specimen came from Cambrian rocks of the Missouri Ozarks and may have formed in association with hot-water emitted from "cracks" in the earth—cracks that later became a series of major faults in the eastern Ozarks. (Value range F.)

"**Red-Earth-Stage**" of the Earth. This iron mine in Minnesota (Mesabi Range) is mining very iron rich rock. The iron was concentrated in the oceans 2.3 billion years ago as a consequence of atmospheric evolution associated with earth's acquisition of an elemental oxygen-containing-atmosphere. Red iron oxide, precipitated from the presence of free oxygen (free oxygen generated by photosynthesis of cyanobacteria and photosynthetic bacteria), accumulated to form these iron-rich deposits now mined to supply the steel used by modern civilization.

Digitate (finger-like) stromatolites, 2.3 billion year old Biwabik Iron Formation, northern Minnesota.

Tillite: these angular blocks represent deposits from ancient glaciers representative of the "**White-Earth-Stage**" of earth's planetary evolution. Some seven hundred million years ago, much of the planet was covered with ice as massive glaciers covered large portions of the globe. Some even suggest that large parts of the oceans froze solid to form what has been called "Snowball Earth."

Close-up of BIF (Banded Iron Formation), the rock formed as a consequence of the earth's atmospheric evolution. This strata was deposited 2.3 billion years ago during the "**Red-Earth-Stage**" of the planet's evolution. Marquette, Michigan.

Ediacarian fossils or Vendozoans: fulfilling the nineteenth century prediction that "complex" life forms equal to trilobites would be found in rock strata precedent to that of the Cambrian. These peculiar fossils were discovered in the mid-twentieth century. Coming from the late Precambrian (Late Neoproterozoic), they represent a major evolutionary step in the development of life, leathery organisms believed by many paleontologists to be unrelated to anything living today or for that matter even ones living in the Cambrian. The stressful conditions of Snowball Earth are believed to have favored the evolution of various multi-celled life forms, which included not only those of the Cambrian but even earlier life like these vendozoans. These organisms appear to have been an evolutionary dead end, going extinct at the end of the Precambrian (Proterozoic Era), some 550 million years ago.

"Green-Earth-Stage:" the last evolutionary stage of the earth, which included the Carboniferous coal swamps. Vast amounts of vegetation accumulated over a large portion of the planet's continents in the late Paleozoic, forming what would become coal beds. The plants of these forests form a major portion of this book!

Early twentieth century version of the Carboniferous (Pennsylvanian) coal swamps. The plants making up the chapters of this book are represented here from left to right. ***Arthrophytes: Chapter 6, Lycopodophytes: Chapter 5, Cordaites*** (strap-like-leaved plant)***: Chapter 7, Ferns and Seed ferns*** (Pteridophytes): ***Chapter 4.***

Late nineteenth century reconstruction (and still valid today) of Permian landscape swarming with many amphibians. Another view of the "Green Earth Stage" is at the end of the Paleozoic Era.

Permian reconstruction: amphibians and early reptiles dominate animal life. Plants were still those of the late Paleozoic; Lepidodendrons (Lycopodophytes), tree ferns, cordaites (left), and arthrophytes-like Calamites (right). The end of the Permia, which was the earth's largest extinction event, would be followed by the Mesozoic, the age of reptiles, including of course the **dinosaurs**.

Permian tree-fern frond: the last of the Paleozoic coal-swamp-plants lived during the Permian Period.

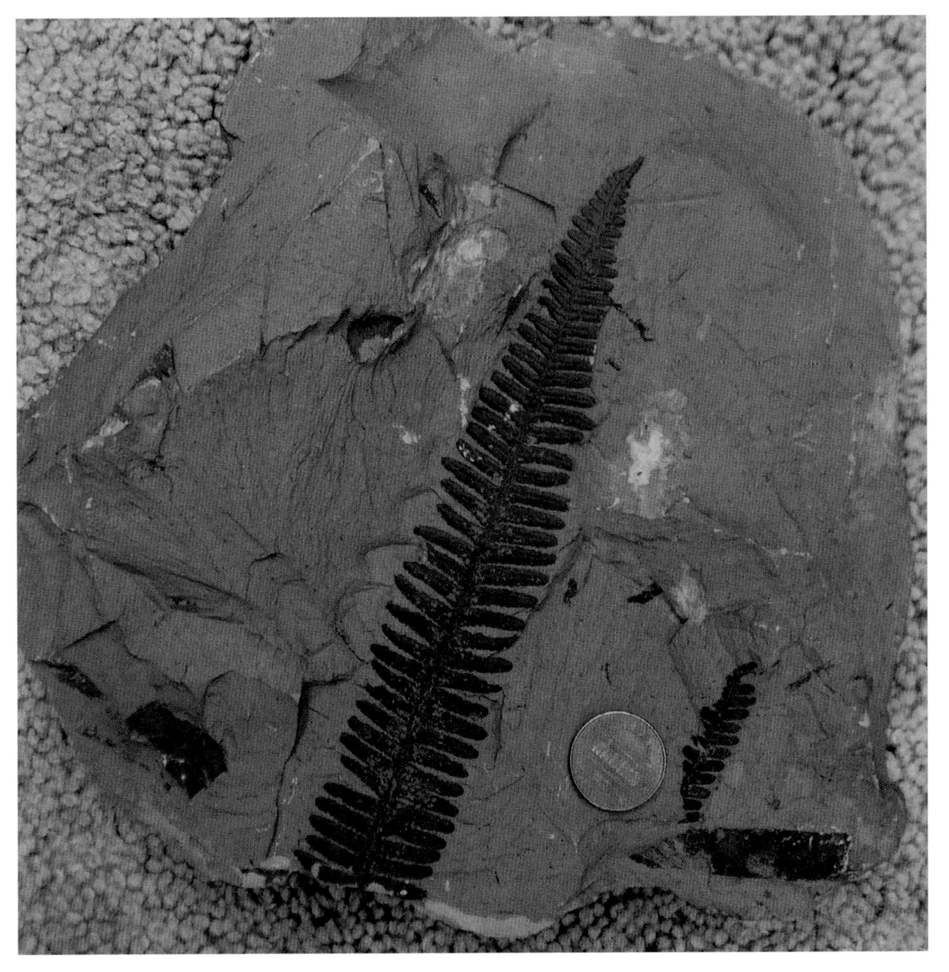

Typical Late Carboniferous (Pennsylvanian) fern fossil: the plant material has been converted to a thin film of coal. This is a compression fossil.

Coal

When land plants became abundant on the earth's surface during the later part of the Paleozoic Era, plant matter could accumulate for tens of thousands of years as a type of sediment. This accumulation of plant material would then undergo compaction and chemical change, the resulting end product being beds of coal, a sedimentary rock composed originally of plants. Coal beds, or coal-like carbonaceous strata, are found throughout geologic time, even in the Precambrian, but it is only in the late Paleozoic that coal becomes thick and relatively free of inorganic matter to be a significant source of energy, a fossil fuel.

Coal seam in eastern Tennessee: the lower part of the seam is at the level of the man's head. The seam was mined prior to making the road cut, which later intersected it.

Thin coal seam (middle) in a fresh road cut, eastern Tennessee: individual coal beds of Pennsylvanian age cover large portions of eastern North America. Generally these are inter-layered with marine strata to form what are known as cyclothems. Coal beds represent the terrestrial (or land portion of the cyclothem), **a time of lowered sea level**. Limestone with its marine fossils represents the marine or high-sea-level portion of the cyclothem. Coal beds (or seams) and associated shale represent the non-marine part. Pennsylvanian (and Permian) cyclothems represent cyclic changes in sea level believed to have been produced by the waxing and waning of glaciers in polar regions of the southern hemisphere, glaciers that existed before Pangaea broke up during the Mesozoic to form the present continents.

The top of a mineable coal bed (Crowberg Coal) exposed during strip mining in western Missouri (1962.)

Coal "smut" layer in Mid-Pennsylvanian strata exposed in a creek, St. Louis County, Missouri. A coal seam can vary in thickness from almost nothing (like this) to being tens of feet in thickness. Generally Pennsylvanian coal seams in the U.S. are thickest to the east, becoming thinner the farther west one goes.

Sandstone beds underlain by shale: both beds have been removed to mine a coal seam in western Missouri. Black piles in the background are mined coal.

Top layer of a coal bed is exposed by removal of overlying strata, using large shovels. People (rockhounds) are collecting pyritized marine fossils embedded at the top of the coal seam, 1958.

Geology field trip to coal strip mine in southern Illinois, 1962.

Think Coal! The author is second to the left with side-arm. Western Alaska, 1982.

Drag line used in removing coal overburden.

Large shovel used in strip mining to remove strata covering a coal seam.

Petrified Wood

If buried in sediment, wood can have its cellulose replaced or charged with minerals, sometimes such replacement being done on a molecular level. This process, known as permineralization, can reproduce all of the wood's original detail, such as grain, rings, and branches with inorganic material that is usually silica (quartz). Most petrified wood found worldwide is of either Mesozoic or Cenozoic age—the wood of conifers or angiosperms of this time especially being subject to permineralization. Paleozoic plants by contrast, because of their reedy or flimsy nature, are less commonly found as petrified wood. In other words, Paleozoic petrified wood is much less common than is that of later geologic time. Related to petrified wood also are the fossil plants found in what are known as coal balls. These are calcareous concretions composed of calcium carbonate, which impregnated the plant material of a coal swamp before the coalification process took place. Like petrified wood, fossil plants preserved in coal balls generally preserve in great fidelity a plants structural detail, which can include cells. Coal balls are especially prevalent where coal seams have been covered by marine sediments, which was the source of the calcium carbonate. Concentrations of coal balls are often found on the craton, as in the U.S. Midwest, where cyclothems with their alternate marine and non-marine sequences occur.

Chunk of late Paleozoic (Pennsylvanian) petrified wood in creek bed near Fulton, Missouri. Paleozoic petrified wood is relatively rare. The punky and reed-like nature of Paleozoic plants usually did not lend themselves to the precise molecular replacement by silica, which characterizes petrified wood. Most petrified wood is either Mesozoic or Cenozoic in age. This wood appears to be from a giant, gnarly tree fern that grew locally in swamps.

Coal ball with *Psaronius* (tree fern) roots. Coal balls represent "petrified" portions of a coal swamp, which can preserve plant structure in detail (often at a cellular level). Coal ball formation is a process which took place shortly after the plant material accumulated. Petrifaction has to take place before the coalification process takes place as coalification destroys most plant structure. A coal seam, unless it contains coal balls, is usually a lousy medium for preserving fossils.

Petrified log, Yellowstone Park, Wyoming: this upright trunk of an early Cenozoic tree has been replaced with silica (petrified or permineralized). The wood was replaced on a molecular level by mineral matter (silica). Paleozoic plants, unlike later ones like this, rarely are preserved in this manner, generally they were too pithy and punky and decayed before permineralization could occur. Late Paleozoic plants have been found (as at Joggins, Nova Scotia) preserved upright like this, but their pithy material was not replaced by minerals as was the wood of this more modern type of tree (angiosperm or conifer).

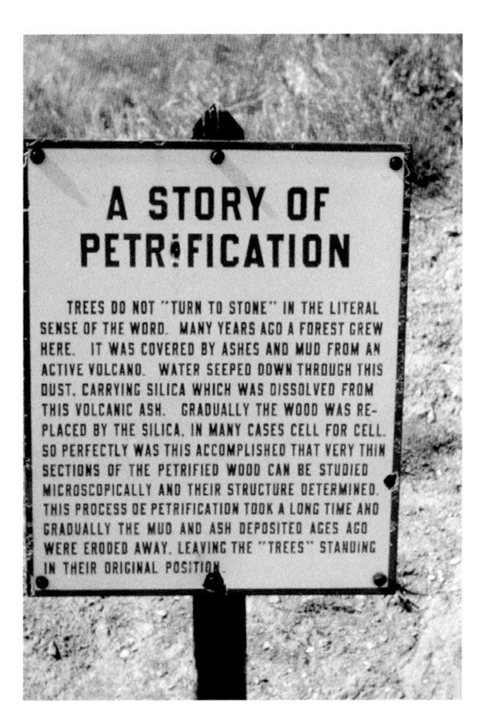

A Story of Petrifaction-sign, Yellowstone Park, 1970.

Ancient Stream Channel Deposits

Fossil plants are often associated with sedimentary rocks like sandstone, siltstone, and conglomerate—rocks that originally were sediments deposited by ancient rivers. Sometimes log impressions or petrified wood is found in these rocks—ancient driftwood carried into what originally were river or stream channels. These ancient streams often drained into subsiding portions of the earth's crust, areas referred to as trenches, troughs, or geosynclines. Portions of the Appalachians and Ouachita Mountains of the eastern U.S. represent such subsiding areas, sediments deposited in them sometimes containing fossils that originally were waterlogged trunks of plants.

Gradational Aspects in the Morphology of Paleozoic Plants

When one actually deals with Paleozoic plants, especially those of the prolific Pennsylvanian floras, there is a peculiarity not seen with younger plants concerning their identification. This peculiarity is that **most Paleozoic plants seem to sort of grade into each other**. With modern plants distinct categories exist. The gaps between these categories usually **not being** occupied by living plants. With Paleozoic floras such gaps are usually not evident. One plant type seemingly grades into another. This taxonomic gradualism often makes it difficult to "nail down" a particular plant fossil as to its specific genus or species. Rather, the plant has to be placed into what is a continuum of morphologies, the best fit being with that species it best matches. With Paleozoic plants, transitory forms are found not only between species but also between genera as well as with higher taxa. Transitory forms can be found even between divisions. Thus, gradational forms are found not only between calamites and lycopods, but also between fern-like taxa (pteridophytes and pteridosperms). This lack of clear morphologic distinctions in the taxonomy of early life has also been noted in animals, especially those of the Cambrian Period, the earliest period of the Paleozoic Era. The phenomena has been attributed to the fact that these organisms are either stem groups or are close to stem groups. Stem groups are forms that had not yet diversified enough from each other that links between them had met with extinction. Extinctions, in later geologic time, would produce the clear morphologic differences separating most major groups of organisms living today.

In other words, the fossil plants that dwelt here were close enough to the beginning of the "Green Earth" stage of planetary evolution so that the clear morphologic distinctions separating plants of today had not yet occurred.

Fossil Plant Form Genera

Plant fossils are almost always found as fragments. Complete plants, especially large ones, are rare. As a consequence of this, different portions of the same plant-like leaves, stems, and seeds are usually given specific Linnaean names. Such names are referred to as form genera and species. They are the norm with paleobotany. Thus, the fern-like leaves of a tree fern are given the generic name *Pecopteris* while impressions of the tree fern's trunk are given the name *Knorria* sp. Fronds of *Pecopteris* are found bearing sori, small clusters of spores born on the underside of *Pecopteris* foliage. Fossil seeds are also found associated with these same tree ferns. These peculiar seeds were given another form genus known as *Trigonocarpon*. In the 1940s, a rare find was made, these seeds were discovered borne on a plant with the fern-like foliage of *Pecopteris*. The "new" plant bearing these seeds was given the name of *Aleothopteris*, but **without the presence of seeds, its foliage is impossible to distinguish from that of** Pecopteris. Such a variety of names

Ancient (Lower Devonian) stream channel deposit, Beartooth Butte, Absaroka Range, Wyoming: the stream channel deposit is the pronounced reddish area near the middle of the butte. Rocks found here contain compression fossils of psilophytes, primitive leafless plants as well as a fauna of bony armor fish (ostracoderms).

Another view of the Lower Devonian stream channel of Beartooth Butte, Wyoming: the overlying beds are Mississippian in age (Madison Limestone); underlying strata is Upper Ordovician and below that occur Cambrian age strata.

Sandstone deposited in river channel deposits (Warrensburg-Moberly Sandstone) exposed during the rerouting of I-70 in north St. Louis County, Mo., in 2002. These thick beds of sandstone were deposited in a channel by an ancient river during the middle part of the Pennsylvanian Period. The black layer at the top is the previous shoulder of I-70 and is not rock strata.

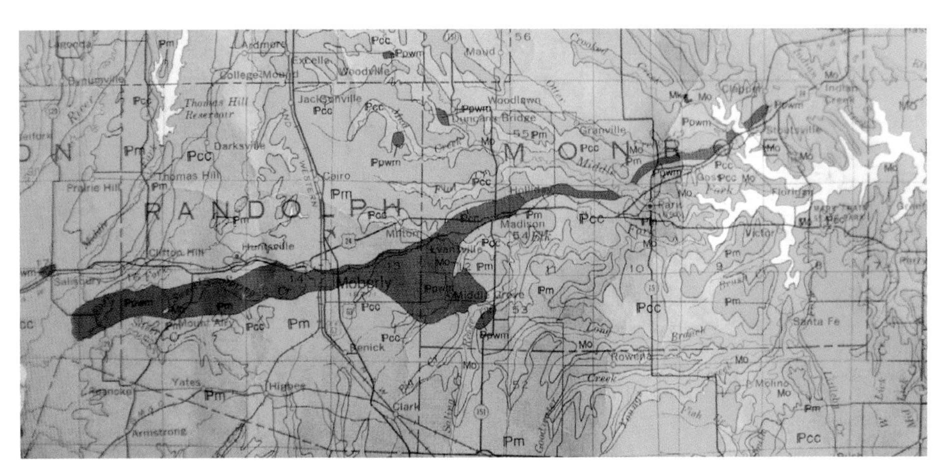

Pennsylvanian age river channel deposits (Warrensburg-Moberly Sandstone) on the 1979 Missouri Geological Map. These channels were cut into older strata by ancient rivers, which deposited sand and clay (sometimes containing washed-in fossil plants).

Another view of these massive river channel sandstones.

Lower beds of the Warrensburg-Moberly channel deposits containing impressions of fossil drift wood (note the impression above the head of the man walking away). I-70 re-routing, St. Louis County, 2002.

Fossil driftwood-filled pocket in Warrensburg-Moberly Pennsylvanian deposits, St. Louis County, Missouri.

Fossil driftwood-filled pocket.

Another view of the fossil plant pocket.

Fossil, rock, and mineral fairs: fossil plants (including Paleozoic ones) often turn up at such fairs, which can be found in most larger communities of the U.S. occurring a few times each year. Often a wide variety of excellent specimens turn up, especially when older collections, containing specimens from classic localities, are recycled. This is a portion of MAPS EXPO, a "fossils only" show held annually in Macomb, Illinois.

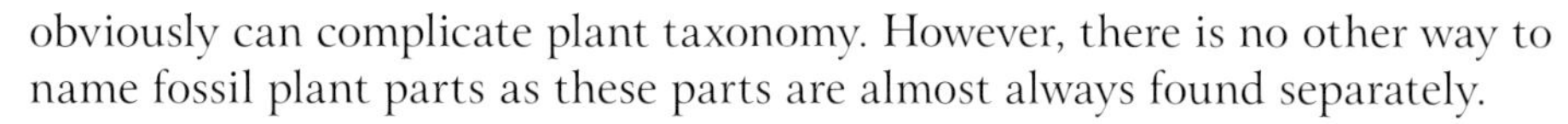

obviously can complicate plant taxonomy. However, there is no other way to name fossil plant parts as these parts are almost always found separately.

Dipping (tilted) Pennsylvanian (Upper Carboniferous) strata, Joggins, Nova Scotia.

Value of Fossil Plants

Considering the amounts of coal mined throughout the world, and the fact that excellent fossil plants **may** occur (but not always) in shale beds overlying such coal seams, this should translate to large numbers of fossil plants originating from coal mining. To some extent this is true. Various impediments to what essentially is fossil salvage, however, may arise from the following:

- Unawareness of the presence of such fossils.
- Coal mining operations being uncooperative toward collectors for various reasons, including, concern for tort liability if someone involved with fossil collecting was to be injured.
- Modern, large scale surface coal mining operations usually quickly bury shale beds overlying the coal seams being mined. These beds are the ones that can contain fossil plants.
- Reclamation requirements require that mined areas be quickly reclaimed and contoured as close to the contour of the original land as possible.

Thick plant bearing sandstones exposed during low tide at Joggins, Nova Scotia, Canada.

Other impediments can include a contempt for collectors by a few (but not all) museum professionals. The author once heard from a group of "professionals" that "all collectors do is make it more difficult for science to acquire fossils." The fact is, however, that most fossils in formal institutional collections have come about as a consequence of collectors. On the "other side of the coin" are those collectors who "squirrel" away significant specimens, never preparing materials or utilizing them, which means never learning from what they collect. Excellent fossil plants do have a commercial value, but what a mindful and serious collector remembers is that scientific and educational values trump commercial considerations.

Ancient filled sinkhole (filled with Pennsylvanian sediments) exposed on I-44 just after it was constructed in 1956. These "filled sink" deposits of the Ozarks have produced elements of an upland flora that are rare compared to plants of the more common coal swamp flora.

Value Ranges of Fossil Plants

Fossil plants often have a lower value on the fossil market than do animals. As a consequence, many are available at low prices and problems associated with excessive commercialization are less prevalent. The value range for fossils used in this book is as follows:

A	$1,000–2,000
B	$500–1,000
C	$250–500
D	$100–250
E	$50–100
F	$25–50
G	$10–25
H	$1–10

Bibliography

Catalani, Johm and Chris Cozart, Ed. *The Pennsylvanian. MAPS Digest*, Vol. 35, No. 2.

Gould Stephen J., 1980. "Bathybius and Eozoon" **in** *The Panda's Thumb*. W. W. Norton County, New York and London.

Stinchcomb, Bruce L., 2012 (abstract). *Conophyton* associated with possible hot spring deposits in the Cambrian of Missouri. Missouri Academy of Science.

Glossary

Biogenicity: the determination that a fossil or fossil-like object was really made by a living thing, rather than formed by some inorganic phenomena like crystallization, metamorphism, or other physical process. In reference to Eozoon, which—until the end of the nineteenth century—saw a major argument within the scientific community as to whether or not it was an actual fossil.

Carbonaceous material: coal-like material usually derived from plant sources. In earthly rocks, carbonaceous material comes from living sources. With extraterrestrial rocks, viz. meteorites, carbonaceous material present is of non-biogenic origin. It appears to have been produced by non-living chemical phenomena in nebulae.

Compression: a common method by which plant fossils are preserved. Compressions consist of some of the original plant material preserved as a coal or lignite film embedded or **compressed** in sediment that has now become rock.

Craton: the stable interior portion of a continent. Mentioned here, craton references coal seams of the U.S. Midwest, which (unlike coal seams at the edge of the craton), were covered by marine waters that deposited limestone (calcium carbonate) above the coal. Sometimes this limestone infiltrated the underlying coal seam to form coal balls before the process of coalification had taken place, thus preserving (in great detail) those plants that were infiltrated by the calcium carbonate.

Cyclothem: a sequence of strata consisting of half marine and half non-marine sediments, which is characteristic of the Pennsylvanian Period of North America. The non-marine portion of the cyclothem usually includes a coal bed. The marine portion of the cyclothem often includes thin limestone beds.

Eozoon Canadense: the first scientifically described "fossil" from some of the earth's oldest rocks, those of the Precambrian. Eozoon turned out to be a "false alarm," a pseudofossil, but it set the stage for discovery of other fossils, like stromatolites, which really are "relics" representative of the earth's earliest life.

Pangaea: the "supercontinent" that existed during the late Paleozoic before the earth's crust "cracked" and expanded. The resulting fragments produced the continents of today. Pangaea is a part of continental drift, which in turn is a part of plate tectonics.

Permineralization: the process by which organic material, such as wood, becomes replaced on a molecule-for-molecule basis with inorganic material, especially quartz. Permineralization is also known as petrification.

Petrification: the replacement of organic material (usually wood) by mineral material, like quartz or limonite. This can be done on an almost molecular basis so that the cell structure of the wood is preserved.

Stem Group: In cladistics the ancestral group that later diversified into a major and distinctive group of related organisms (clade). Stem groups may remain pretty much the same over a span of geologic time then undergo taxonomic radiation with a burst of diversity

Chapter Two
"Algae" and Fossil Marine Plants
An Overview of Fossil Plants through the Paleozoic Era

As far as is known, land vegetation did not exist prior to some 400 million years ago, that is prior to the Silurian Period of the Paleozoic Era. Late Precambrian time (and into the Cambrian as well) may have seen some form of algae living in damp, shaded areas, **but there is no clear fossil evidence for this!** ***Late Precambrian land-deposited-sediments (terrestrial deposits) lack any evidence of obvious life, although its presence has been deduced questionably through geochemical means, analyzing ancient (fossil) soil layers.***

Unquestioned land vegetation appears to be lacking in the Cambrian. It has questionably been found in the Ordovician and more distinctly in the Silurian as small (dichotomizing) fossil plants. These are attributed to something related to psilophytes—primitive leafless land plants. Silurian strata has also yielded enigmatic lycopod-like plants that may (or may not) be related to the lepidodendrons of the late Paleozoic, the subject of Chapter Five. The first undoubted and widespread land plants appear in the Devonian Period. These are ferns and tree ferns of the Divisions Pteridophyta (true ferns) and Pteridospermophyta (extinct seed ferns) are the earliest easily recognizable fossil land plants. Recognizable plant foliage of this time includes the ubiquitous *Archaeopteris* and large fossil logs (usually sediment-filled casts) given the name of *Eospermatopteris*.

Fossil land plants of the Lower Carboniferous (or Mississippian Period) include, more often than not, a variety of lycopods. Lycopods and related forms are the subject of Chapter Five. This is in contrast to the Upper Carboniferous (Pennsylvanian Period of North America) where the various divisions of Paleozoic plants reach their zenith (and to which a chapter to each division is devoted in this book).

The Upper Carboniferous (or coal age) flora, because of its widespread occurrence over the northern hemisphere, is the best known flora of the Paleozoic Era. This is, in part, because of its association with mineable coal deposits over this vast region. Upland floras of the same age, in contrast, are more poorly known. Most of the "coal swamp" flora is early or middle Pennsylvanian in age. Pennsylvanian floras from near the end of the period are different from those of the coal swamps, reflecting, in addition to evolutionary changes, a change in the world's climate, which is headed toward the drier climate of the Permian Period. The profusion of plant life in the Pennsylvanian may have increased the planet's atmospheric free oxygen content to a level above the twenty-one percent that exists today and may have affected Permian ecosystems.

Floras of the northern hemisphere's Permian Period, while still exhibiting many elements of the "coal age" and its swamps, included some new and unusual elements, these often being very localized. Floras of the southern hemisphere are also different and are represented by the *Glossopteris* flora

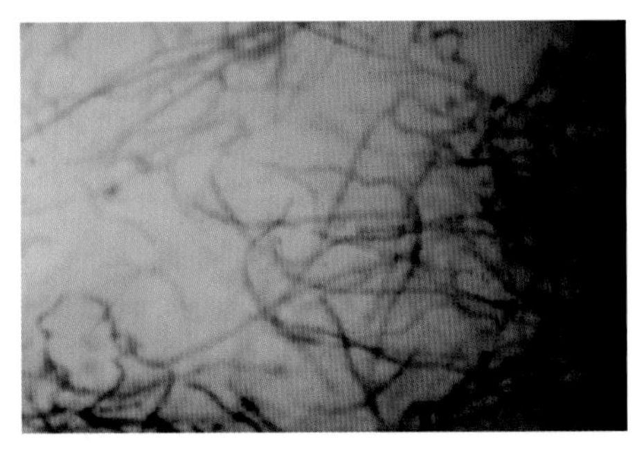

Cyanobacterial filaments viewed through a microscope. The blue-green-algae or cyanobacteria were considered to be primitive plants until the 1980s. They have a (somewhat) extensive fossil record where they formed the stony-masses known as stromatolites. Cyanobacteria represent the oldest photosynthetic organisms, which have left a relatively uncontroversial fossil record. They, however, are no longer considered to be plants.

(the subject of Chapter Eight). The *Glossopteris* flora is believed to be that of a more temperate climate, one associated with cooler temperatures associated with glaciation.

The Most Primitive of the Primitive: Blue-green "Algae" and Stromatolites

The topic of this chapter, non-vascular plants, prior to the 1980s would have included the blue-green algae, the fungi, and even the bacteria. It is now known that both bacteria and blue-green algae (now called **cyanobacteria)** possess a cell type that lacks a cell nucleus (prokaryotic cell). Because of this, "blue-greens" are now placed in their own kingdom, the kingdom Monera.

Surface of an algal (cyanobacterial) mat: associated with a mass of cyanobacterial filaments in such a mat is a considerable amount of slime so that "algae" growing on rock surfaces in shallow water can be very slippery.

Stromatolites: these stony domes were produced by cyanobacteria growing in a shallow sea about half a billion years ago. This is a small exposure of what would be a sizeable stromatolite reef if the grass and soil shown at the top of the photo were removed.

Font covered by a lush mat of cyanobacteria, Hot Springs National Park, Arkansas. Cyanobacteria can thrive on a wet, sun-lit surface like this.

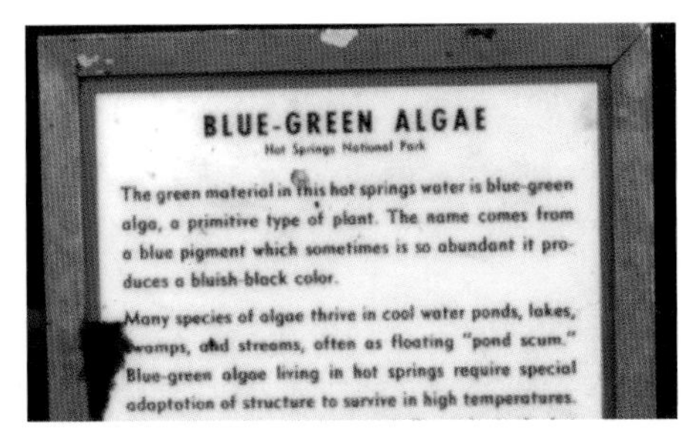

Blue-green algae information: blue-green algae, however, **are no longer considered plants**. They are now placed in the Kingdom Monera, along with bacteria and archeobacteria.

A single large stromatolite: stromatolites, or stroms, come in a variety of shapes and sizes. This one, if sliced longitudinally, would resemble a slice through a cabbage. It has a depression in its middle that holds water. Cambrian, Fredericktown, S E Missouri. (Value range F.)

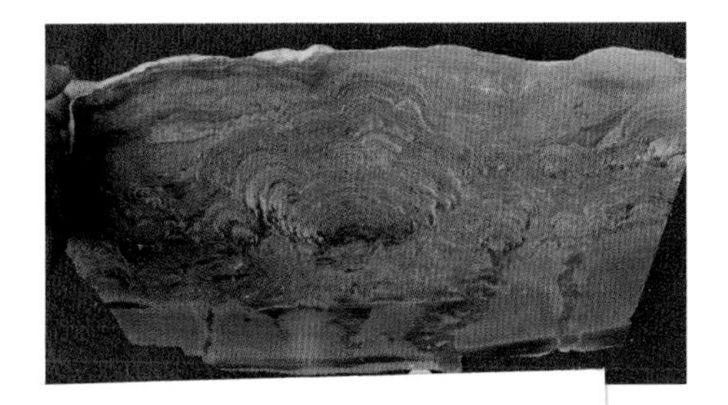

Longitudinal cross-section through a stromatolite: the reddish color of the stromatolite domes occurs because they contain oxidized (ferric) iron—red iron oxide formed from elemental oxygen given off by photosynthesis of (once-existing) blue-green algae. Green sediment around and under the "strom" is from less oxidized ferrous iron present in sediments that were not affected by the presence of free oxygen. Mid-Proterozoic, Belt Series, NW Montana. (Value range E.)

Glaciated surface of a 2.3 billion year old stromatolite reef. Stromatolites are especially well developed prior to six hundred million years ago as the various types of grazers (like small marine animals) had not yet appeared to feed upon what might be considered as giant cyanobacterial salads. Stromatolites almost disappear from the rock record 430 million years ago as many organisms appeared at this time that could feed on them. These organisms were cropping them and thus preventing them from forming.

...se are giant stromatolites formed on the ocean floor some ... They have been tilted at a sixty degree angle by ancient ...ock Lake Iron Mine, Atikokan, Ontario, 1985.

Stromatolites and Heavy Element Mineralization

Stromatolite bearing strata (mainly limestone and dolomite) can be the host rock for mineral deposits, especially zinc, lead, and copper minerals. The porosity of ancient stromatolite reefs, along with residual organic matter from the stromatolites, may have influenced and encouraged mineral deposition. Such algal or stromatolite mineral depositions are associated with horizontal beds of limestone and dolomite, which are known as stratiform mineral deposits.

... from the Bulawayan Group, Bulawayan Greenstone Belt, Zimbabwe, South Africa. The lamina seen here have a signature characteristic of a stromatolite, although no fossil filaments are observable under magnification. The overall structure, however, is characteristic of a stromatolite. It has the distinctive stromatolite signature. Stromatolites from Zimbabwe are some of the oldest stromatolites and thus are some of the **oldest known fossils**.

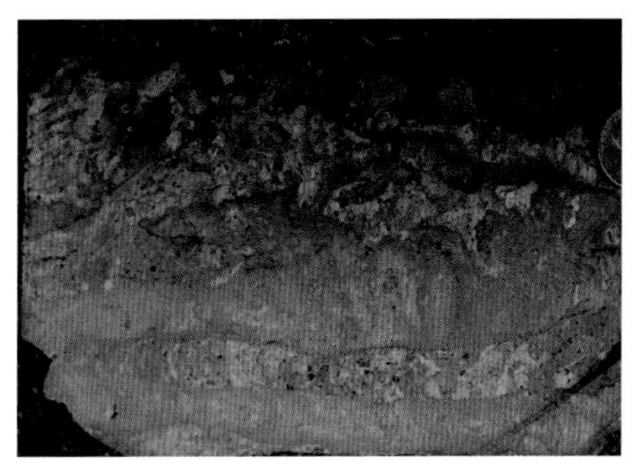

Collenia sp. Digitate stromatolite associated with lead mineralization (galena or lead sulfide) from Missouri's new lead belt (Viburnum trend). Stromatolite reefs have acted here as a recipient for lead mineralization.

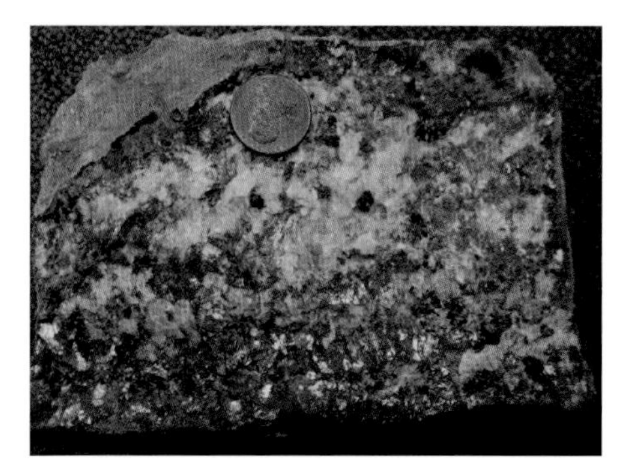

Lead ore (disseminated galena) from a mineralized stromatolite reef zone in Missouri's "lead belt."

Galena (lead sulfide) crystals from the Viburnum Trend (Missouri's new lead belt).

Rich galena mineralization in Brushy Creek Mine, Viburnum Trend. The richest zones of galena mineralization are often those associated with Cambrian stromatolite reefs in both the old and new lead belts of Missouri.

Hand digging galena and barite in the Washington County district, which is in Missouri's lead belts. Stromatolites are associated with the mineralized layers.

Quartz mass in Washington County's lead-barite-district formed from digitate stromatolites (stromatolite "fingers"). As in Missouri's lead belts, stromatolites in some way have also influenced mineralization in these deposits.

Similar to the previous quartz replaced digitate stromatolites, Tiff, Missouri, Washington County. The green growth at the far left is moss, a small non-vascular living land plant that has recently grown on this stromatolite mass.

Bacteria

Bacteria represent a large group of monerans that, before their prokaryotic cell-type was appreciated, generally were placed in the Plant Kingdom. Bacteria can form stromatolite-like structures, but these are rarer than stromatolites formed by cyanobacteria. Photosynthetic bacteria, along with cyanobacteria, may have been responsible for BIF (Banded Iron Formation), a type of sedimentary rock restricted to strata of the early earth.

Girvanella sp.: a group of small stromatolites from the late Pennsylvanian (or early Permian) rocks northwest of Topeka, Kansas. (Value range F.)

Somphospongia sp.: a distinctive type of stromatolite formed (possibly in part) from photosynthetic bacteria. From the Upper Pennsylvanian, Burlingame Limestone, Jackson County, eastern Kansas, (Value range F.)

Conophyton sp.: these cone-in-cone-like stromatolites can be associated with areas of geothermal activity, especially hot springs. This specimen came from Cambrian strata on the Ste. Genevieve Fault System, a major fault in the eastern Ozarks that was active in the Mid-Paleozoic. Heat-loving algae and bacteria, which made this stromatolite, may have lived in association with geothermal activity occurring along this fault before it became tectonically active in the Devonian Period. (Value range F.)

Two stroms given the Linnean name *Cryptozoon* (hidden animal) as these fossils (in the late nineteenth century) were considered as a type of giant foraminifera (a type of protist). Cambrian, Antrim Basalt, Katherine, NT, Australia. (Value range E.)

Conophytum basalticum: a Cambrian stromatolite found in northern Australia, which has been distributed among collectors through the Tucson show. Like *Conophyton* (notice the difference in spelling), *Conophytum* may also have been associated with geothermal activity.

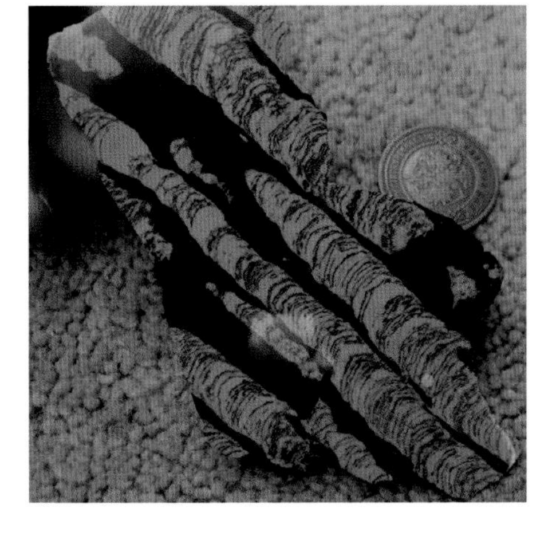

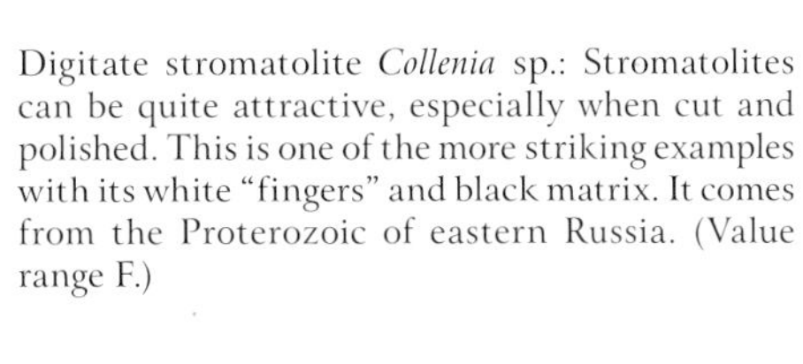

Digitate stromatolite *Collenia* sp.: Stromatolites can be quite attractive, especially when cut and polished. This is one of the more striking examples with its white "fingers" and black matrix. It comes from the Proterozoic of eastern Russia. (Value range F.)

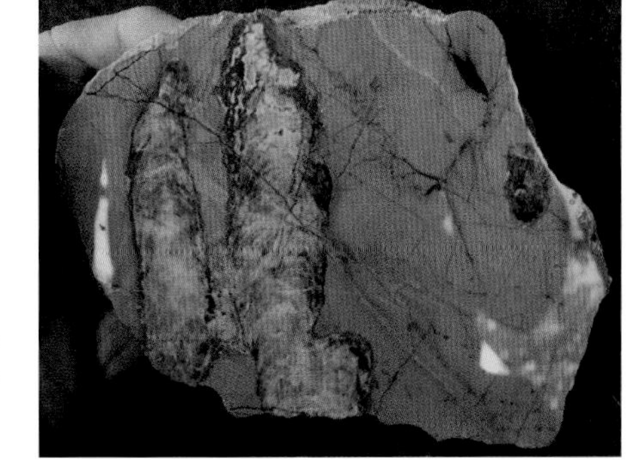

Longitudinal slice of a digitate (finger-like) stromatolite. Proterozoic, Western Australia. (Value range F.)

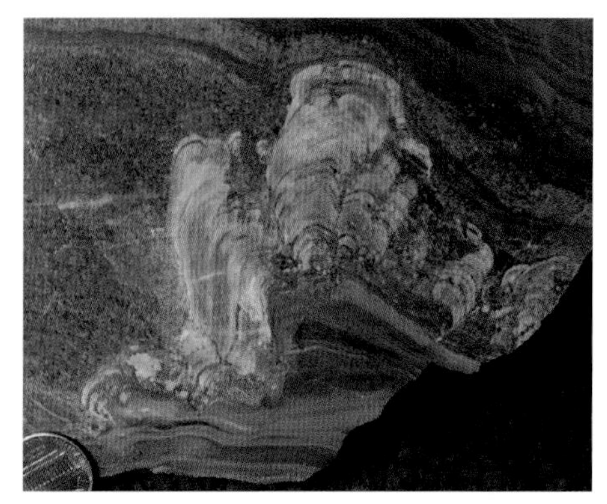

Stromatolite that grew upon layers of a greenish substrate. The red color of the stromatolite and the surrounding matrix comes from iron oxidization. Oxidized iron formed when the strom was growing. This happened because elemental oxygen given off by the photosynthesizing cyanobacteria would oxidize **ferrous** iron oxide (the pigment in the greenish substrate) forming red **ferric** iron oxide. The greenish substrate on which the "strom" grew thus has its iron in the ferrous (less oxidized) state. Mesoproterozoic, Belt Series. Western Montana.

Stromatolite cube: cut from a 1. 5 billion year old stromatolite reef, from what is now Bolivia. A number of these interesting stromatolite "cubes" have come through the Tucson Show. (Value range F.)

Vertical (longitudinal) slice through a "*Cryptozoon*" stromatolite. Cambrian, Upstate New York.

Banded iron formation (BIF): this unique, ancient sedimentary rock is believed to record the transition of the earth's atmosphere from a reducing (oxygen-free) one to an oxygen containing one. BIF is thus a record formed as a consequence of the release of elemental oxygen given off by photosynthesis. The red bands are jasper, a mixture of silica and hematite (ferric oxide). The grey layers are hematite (ferric oxide) precipitated from iron-rich seawater in the presence of photosynthesizing cyanobacteria (or photosynthetic bacteria) in the water column 2.3 billion year ago. Michigamme Iron Formation, Marquette, Michigan.

Iron ore (ferric oxide) as banded iron formation (BIF) exposed in an iron mine in northern Minnesota where hematite, precipitated from iron originally dissolved in sea water. This iron oxide is believed to have been concentrated and precipitated during the earth's acquisition of its oxygen containing atmosphere. That span of geologic time in the Precambrian when this period of BIF formation was prevalent is known as the "Red Earth Stage" of earthly evolution.

Archeobacteria

Another group of monerans, the Archeobacteria, or archaea, are a group of bacteria-like (but genetically distinct) micro-organisms that are the most heat-tolerant life forms to exist—the thermophiles. Found living in the interior of hot springs, these genetically distinct organisms may be the most primitive of life forms and date back to the beginnings of life on earth, a beginning possibly associated with geothermal activity. Archeobacteria not only thrive under the most severe thermal conditions, but other types, like the halophiles, can flourish under hypersaline environments other life forms cannot tolerate.

Fungi, having cells with a cell nucleus (eukaryotic cell type) like true plants (but lacking the ability to photosynthesize), are placed in their own kingdom (the Fungi). Lichens, also previously considered as plants, represent an example of symbiosis—a cooperative arrangement between fungi and cyanobacteria—and hence they also are not plants.

Peculiar "multiple stromatolite," preserved in chert, Cambrian, southern Missouri.

BIF, banded iron formation, from Australia: colorful iron formation like this can create an attractive rock. It is often used in lapidary work and known to rockhounds as jasper or "tiger iron." (Value range F.)

Azurite buttons: azurite is a mineral composed of copper carbonate. These azurite buttons are believed to have been formed by the physiologic activity (life activities) of chemosynthetic bacteria. Bacteria can use a number of energy sources besides sunlight as a source of metabolic energy, one of which is **chemical energy**. Chemosynthetic pathways utilize the chemical energy released in forming heavy metal compounds, like copper minerals. Photosynthesis is thus but one of many energy pathways utilized by organisms of the early earth, organisms known as monerans. **These azurite buttons are believed to be a type of chemotropically formed stromatolite formed by either chemosynthetic bacteria or archeobacteria.** (Value range F.)

Close-up of azurite button (copper carbonate), chemosynthetic stromatolites.

Hot spring (with mud): archeobacteria can live in the inner reaches of a geothermal vent like this. Archeobacteria are believed by microbiologists to represent another kingdom of life separate from bacteria as genetically they are quite different from bacteria (and cyanobacteria).

Hot Springs, Yellowstone Park, Wyoming: Hot Springs supports a variety of monerans from **thermophyllic archeobacteria** (Archaea), which live in the spring's interior, to **chemoautotrophic bacteria**, which live nearer its surface, and **cyanobacteria**, which live in wet areas of the spring branch exposed to sunlight. Geothermal areas like this are believed to be the most primitive and fundamental ecosystem on the planet, an ecosystem going back over 3. 5 billion years.

Geyser vent: geysers, a type of spouting hot spring, are found today in areas of extensive geothermal activity. During the early earth, geothermal activity appears to have been much more active and widespread than it is today. The earth's source of geothermal energy, **radioactivity**, was more intense and widespread in the deep geologic past.

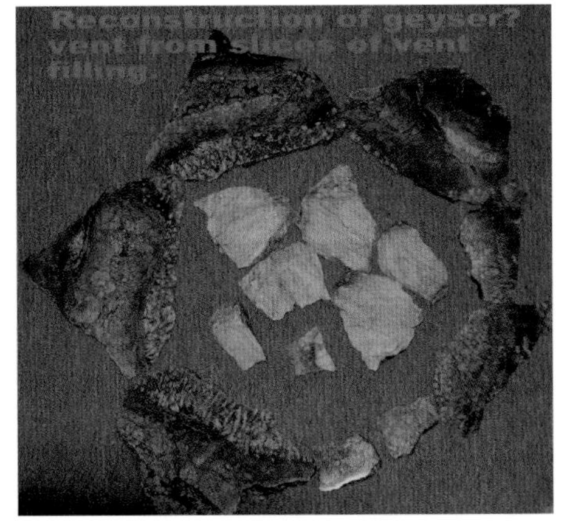

Geyser conduit reconstructed from what is believed to have been an ancient (1.5 billion year old) conduit discovered during the cutting into 1.5 billion year old igneous rock (felsite) in southern Missouri. The center of the conduit was filled with white crystalline calcite. The walls were made up of a complex arrangement of hematite, calcite, pyrite, and dolomite.

Interior of the upper portion of a geyser conduit, Yellowstone Park, Wyoming. Archaea live in the inner reaches of such geothermal environments, environments that would be lethal to other life forms.

Hot spring or geyser vent material with fragmented layers of carbonate minerals formed (possibly) from either thermophyllic bacteria or from archeobacteria. These carbonate layers originally appeared to have formed hollow spheres that may have fragmented from intense seismic activity accompanying the igneous activity. Such igneous activity was intense in mid-Proterozoic, southern Missouri.

Calcareous Red and Brown Algae

The current definition of algae consists of those non-vascular plants that are not mosses, liverworts, or hornworts. This includes the Green-algae, some of which can be similar to cyanobacteria, but which have a **eukaryotic cell type**. Also included here are the diatoms, the brown algae, and the red algae, all of which are found living either in fresh or salt water. Sometimes considered as algae are the Charophytes or stoneworts, interesting plants that live in both fresh and salt water. Of these various plant groups, the **green** and **red algae** (especially the calcareous forms) have left the best fossil record, a record whose fossils usually are found in marine rocks like limestone. The calcareous algae actually have a fairly good fossil record. Many of the different types of marine algae found in the fossil record were previously considered sponges.

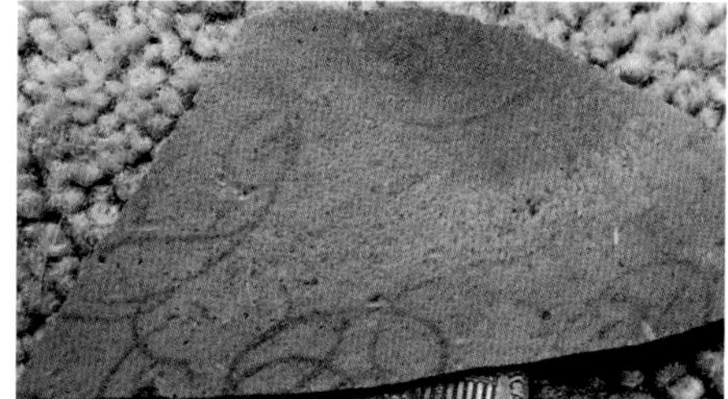

Grypania spiralis: A problematic plant-like fossil from 2.1 billion year old rocks. This is an example of a fossil that is "out of place" with respect to geologic time. If *Grypania* was found in the Paleozoic, it would be considered as a type of sea weed or algae composed of eukaryotic cells. It is many times older, however, and thus should have been composed of prokaryotic (non-nucleate) cells. *Grypania* is possibly an arrangement of monerans to form a type of large organism that lived early in the history of life when most life forms otherwise were very small. Early Proterozoic Negaunee Iron Formation, near Marquette, Michigan (Value range F.)

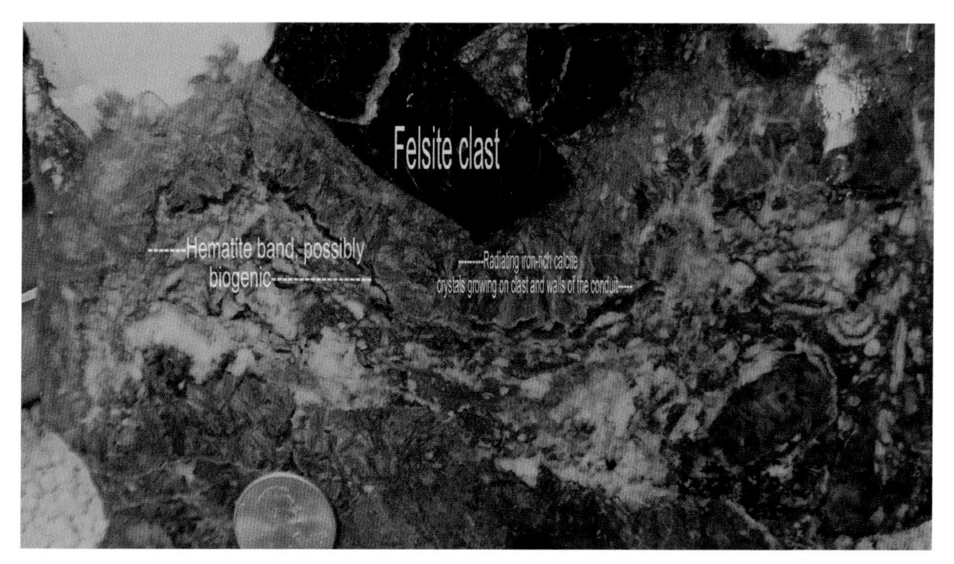

Portion of the wall of a geothermal conduit. Labeled in the image is a layer of iron-rich calcite made up of small domes, possibly of stromatolitic origin.

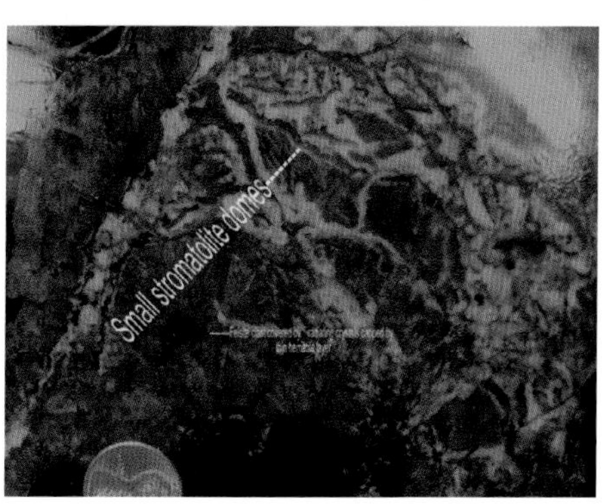

Crystals of iron-bearing calcite that formed on a felsic clast, to the left of that are thin bands of hematite of possible biogenic origin.

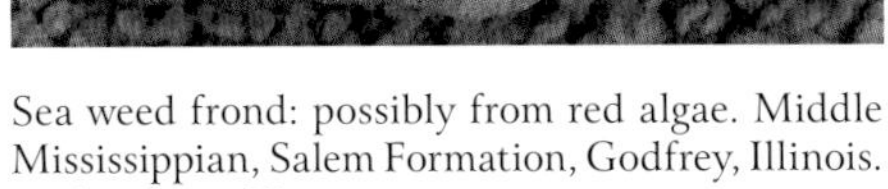

Sea weed frond: possibly from red algae. Middle Mississippian, Salem Formation, Godfrey, Illinois. (Value range F.)

Red(?) Algae: some time over half a billion years ago, eukaryotic cells "clumped together" to form various types of what are conventionally known as sea weeds. Mississippian, St. Louis Limestone, St. Louis County, Missouri. (Value range F.)

Part and counterpart of a red algae frond. Salem Formation, Godfrey, Illinois.

A small "frond" of red algae. Salem Formation, Middle Mississippian, Godfrey, Illinois.

"Leaves" of sea weeds (red algae) scattered on the sea floor.

Portion of a bifurcating red algae stem.

Brown algae: another example of what is probably the sediment-filled cavity of a fucoid. This has a repeating pattern suggestive of a lepidodendron, however, land plants like lepidodendron leave a carbonaceous residue from the cellulose they contain. Marine algae contain no cellulose in contrast to land plants, so they leave what is usually only a vague impression like this. Mississippian, St. Louis Limestone, St. Louis, Missouri.

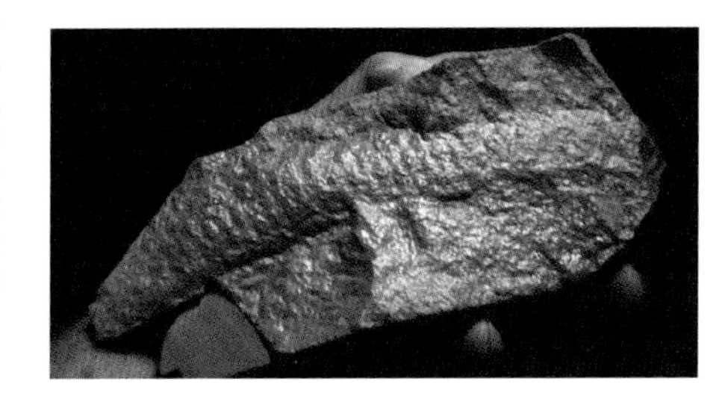

Small spray of red algae.

Branching red algae? Devonian Little Saline Formation, Ozora, Missouri.

Close-up of Cambrian fossil "fucoid(?.)"

Brown algae(?): fucoids, large gelatinous sea weeds, may have produced this fossil. Many so-called fossil fucoids have turned out to be animal (worm) burrows formed on the sea floor or made just below the surface by animals burrowing in sea floor sediments. This specimen may be the sediment-filled interior of a brown algae or a worm burrow. It is often difficult to distinguish between these two types of phenomena with fossils like this. Cambrian, Crawford County, Missouri.

Charophytes

The charophytes or stoneworts are small plants that usually live in fresh water. Their small calcareous oogonia (reproductive structures) can locally be common microfossils, but actual fossil charophyte plants like these are rare.

Living charophytes or stoneworts: these water plants live in fresh water; they are often associated with calcareous springs.

Fossil charophytes: the specimen on the right is the best example of this fresh-water-plant. It somewhat resembles a small specimen of *Equsitum*, an arthrophyte (see Chapter Six). Charophytes represent a division of plants that appeared during the early Paleozoic. Lower Silurian, Pike County, Missouri. (Value range F.)

Charophyte: the plant has left a carbonaceous film, but one that is not as intense as if it were composed of cellulose. Charophytes are delicate plants that produce small calcareous spheres called oogonia. These are found as microfossils much more commonly than are impressions like this. Edgewood Dolomite, Pike County, Missouri.

Calcareous Green Algae

A variety of calcareous fossils often attributed to sponges are known from Paleozoic marine limestone. Many of these are now attributed to either calcareous green or red algae. Some of these occurrences may also be examples of symbiosis, a cooperative arrangement between a photosynthetic algae and a marine animal. Archimedes, a puzzling bryozoan, is believed to be an example of this as are the puzzling archeocyathids.

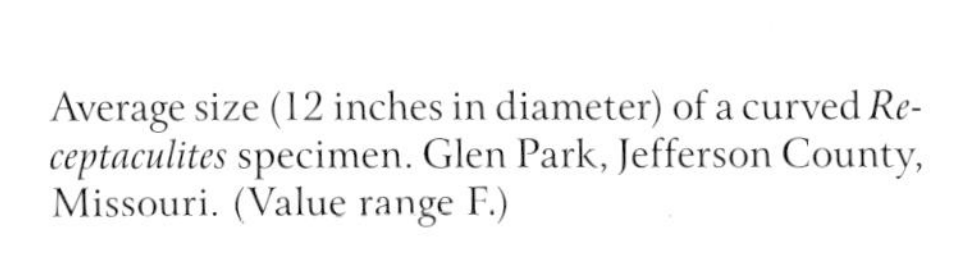

Average size (12 inches in diameter) of a curved *Receptaculites* specimen. Glen Park, Jefferson County, Missouri. (Value range F.)

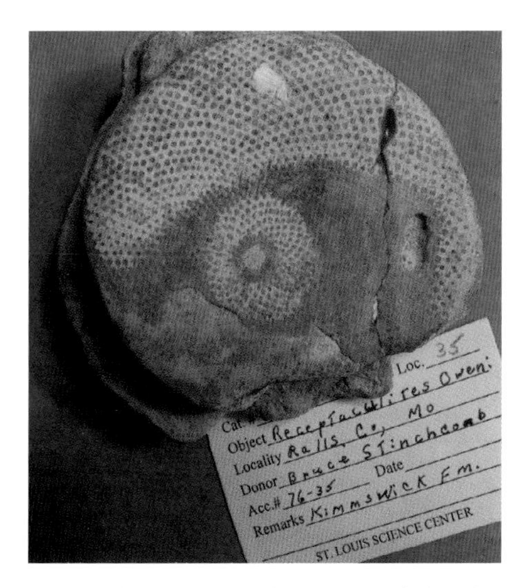

Receptaculites sp.: a small specimen of this distinctive fossil found as a creek pebble. In earlier literature, *Receptaculites* is classified as being a type of sponge, which it is not! Kimmswick Limestone, Ralls County, Missouri. (Value range F.)

Weathered surface of average size, disk-shaped (10 inches in diameter) *Receptaculites*. Kimmswick Limestone, Ralls County, northeast Missouri. (Value range F.)

Weathered surface of small *Receptaculites* disk. Kimmswick Limestone, Jefferson County, Missouri. (Value range G.)

Weathered portion of a curled (or globose) *Receptaculites* specimen. Ralls County, northeastern Missouri. (Value range F.)

Weathered cross-section similar to the previous example. These are very distinctive fossils indicative of the Middle and Late Ordovician. Even a small section of *Receptaculites* is readily identifiable. (Value range G.)

Portion of a small, irregularly-shaped *Receptaculites*. Kimmswick Limestone, Jefferson County, Missouri.

Globular-shaped, silicified *Receptaculites* specimen. Rock Creek, northern Jefferson County, Missouri. Silicified specimens like this, usually showing more detail then specimens in limestone, are desirable fossils. (Value range F.)

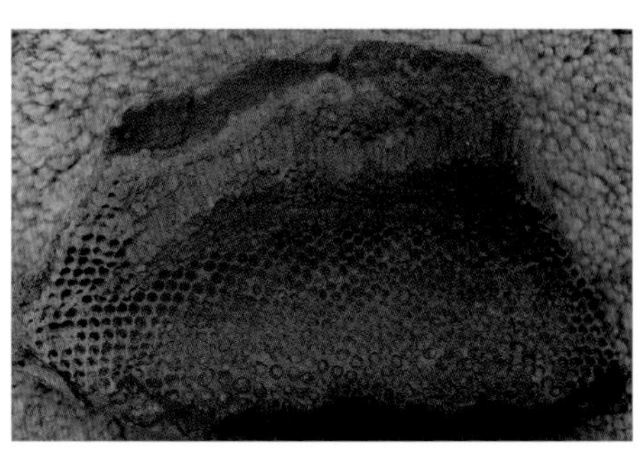

Outer portion of disk-shaped *Receptaculites,* where the "cylinders" have been eroded away. Kimmswick Limestone, Glen Park, Jefferson County, Missouri. (Value range G.)

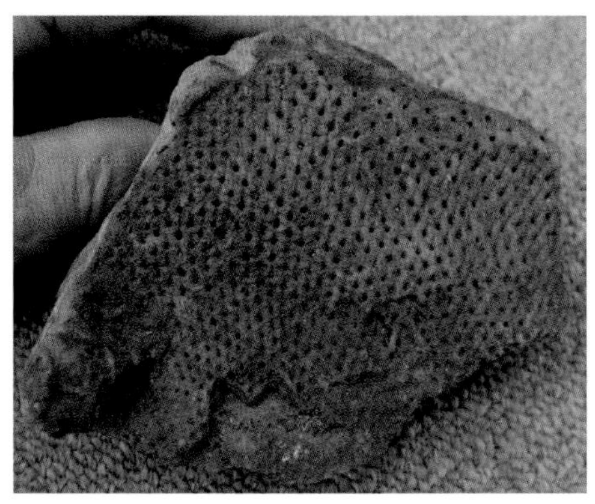

Receptaculites oweni: portion of a large (average size) *Receptaculites* from the Lander Sandstone of Upper Ordovician age. Big Horn Mountains, Wyoming. (Value range F.)

Ischidites sp.: a small receptaculitid with a quartz vein cutting across it. Garden City Formation, Middle Ordovician, Millard County, Utah. (Value range F.)

Section of Cambrian and Upper Ordovician strata exposed in the walls of Shell Canyon, Big Horn Mountains, Wyoming. The Lander Sandstone at the base of the Upper Ordovician strata and the overlying Big Horn Dolomite both yield *Receptaculites oweni* associated with other fossils representative of the arctic Ordovician—strata deposited on the other side (now north) of the transcontinental arch.

Cyclocrinites dactyloides: these golf-ball-like fossils appear to be relatives of *Receptaculites*—that is, like *Receptaculites*, they may also be Dasycladacean calcareous algae. Hopkinton Dolomite, Farmers Creek Member, Jones County, Iowa. (Value range F for group.)

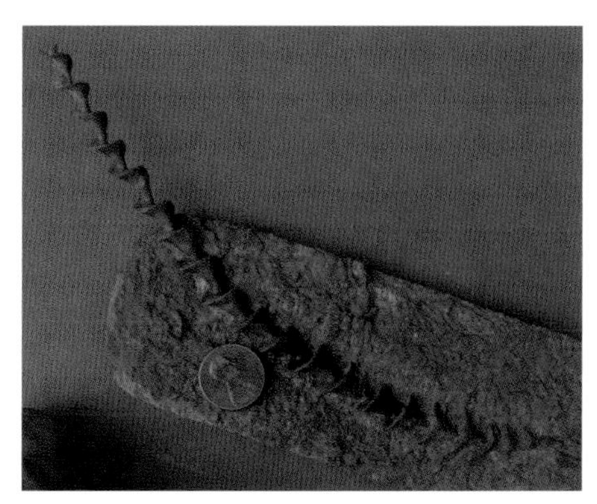

Archimedes wortheni: slender form.

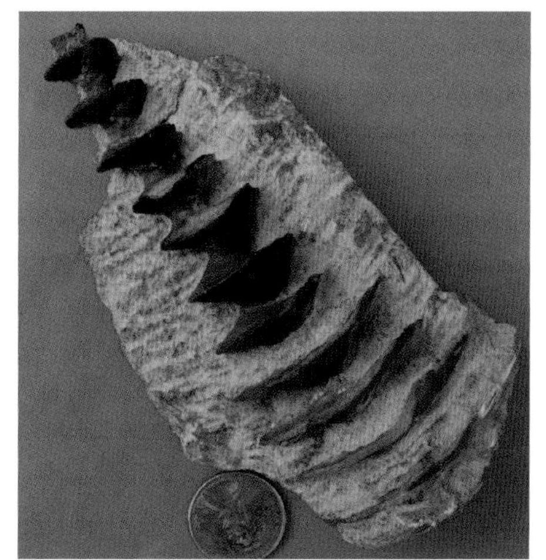

Archimedes wortheni: robust form. (Value range F.)

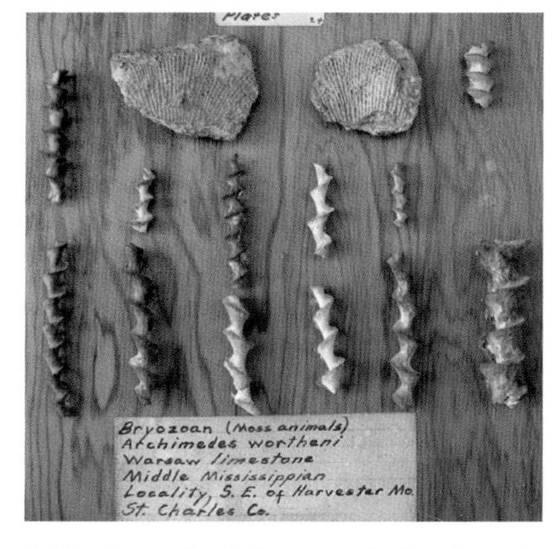

Archimedes wortheni: the genus *Archimedes* is, in part, a bryozoan—an animal phylum of small marine and fresh water lophophorates. *Archimedes* is thought, by some paleontologists, to represent a symbiotic relationship between a bryozoan and a type of calcareous red algae. *Archimedes* is a common and widespread Mississippian fossil in the U.S. Midwest. (Value range F.)

Archimedes communicus: a species of *Archimedes* characteristic of the late or Upper Mississippian. Glen Dean Limestone, Chester Group, Upper Mississippian. (Value range F.)

Archimedes communicus: a smaller species of the genus *Archimedes* characteristic of the Upper Mississippian (Chesterian) strata of the U.S. Midwest and South.

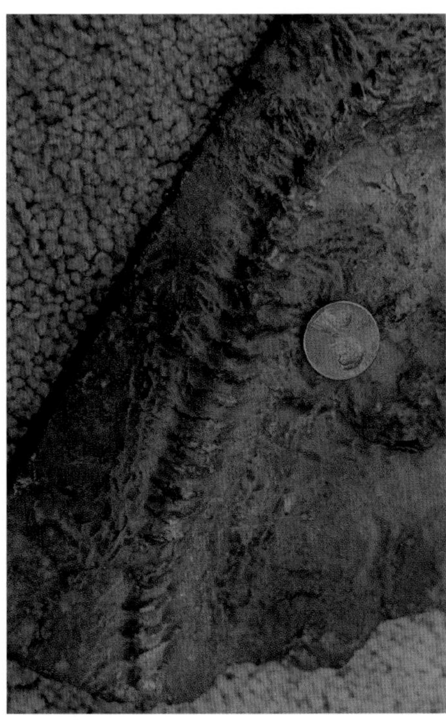

Archimedes sp.: Pitkin Limestone, Jasper, Arkansas.

Archeocyathid, *Cambrocyathus* sp.: a typical archeocyathid, thought by some paleontologists to be a type of calcareous algae. Lower Cambrian, Labrador, Canada.

Prototaxites

This is an example of a problematic fossil! It is locally abundant in Devonian shales of Indiana and Ohio, as well as elsewhere. It resembles a large trunk of petrified wood and was previously considered as such, hence the names ending, *taxities*, after a type of conifer. *Prototaxites* has a peculiar structure unlike that of wood (or petrified wood) and is also found associated with marine strata. Also, no actual trees with extensive woody tissue existed in the Devonian, although there were "tree ferns" (*Archeopteris*) and the pithy trunk of a "tree fern" known as *Eospermopteris*. *Prototaxites* does not have the pithy structure of *Eospermopteris*, the name (genus) given to large and poorly preserved Devonian trunks presumed to be from *Archeopteris*. One interpretation of *Prototaxites* is that it formed from mats of marine liverworts rolled together somewhat in the manner of a giant liverwort-cigar. Why the mats of liverworts would always be rolled together in this manner is puzzling as one would think they would also be preserved as a mat. It has also been explained as representing some type of giant marine fungi that grew in the cigar-like masses on the sea floor. Associated with *Prototaxites* is a puzzling fossil that occurs in vast numbers and is known as *Protosalvinia*. *Protosalvinia* has been suggested as being related to *Prototaxites* in some manner, and again it may represent some type of fungi.

Prototaxites sp.: a Devonian paleontological (and paleobotanical) enigma. Note the fine structure in this sliced (and polished) section of what looks like a fossil log. This is not a structure associated with petrified wood. It is odd!

Bibliography

Andrews Jr., Henry N., 1964. *Ancient Plants and the World They Lived In.* Comstock Publishing Company, Ithaca, New York.

Glossary

Archaebacteria: prokaryotes that are genetically distinct from other bacteria (eubacteria) and are capable of living under conditions too severe for other life forms. The archeobacteria consist of the thermophyles (heat loving forms), the halophyles (salt loving forms), and the methanogens (methane producing forms).

Chemoautotrophic bacteria: use physiological pathways involving chemical reactions, such as oxidization of heavy metals like copper. This is in contrast to photoautotrophic bacteria that, like plants, utilize light in photosynthesis as a source of physiological energy.

Chapter Three
Psilophytes and Other Early Vascular Plants
Living Fossils

Most of the plants represented here can be found growing today. They are referred to as **living fossils**. These are organisms having changed little over geologic time and often still occupy specific ecological niches. Such plants include the liverworts, which—although rare as fossils—have been found in Rhynie Chert, a layer of rock originally part of a Scottish Devonian peat bog that became silicified, thus preserving plants like liverworts, plants that otherwise would normally not leave a fossil record. Psilophytes and mosses also have been found in the Rhynie Chert.

Modern mosses: or bryophytes are small, herbaceous, non-vascular land plants requiring shaded, damp places to live. They are believed to be ancient land plants—actual fossils go back as far as the Lower Devonian, where they have been found in the Rhynie Chert of Scotland. Because of where they live and their delicate nature, they are rare fossils.

Land plants today consist of two major categories of photosynthetic life: **vascular** and **non-vascular plants**. Vascular plants are those possessing vascular bundles (xylem and phloem), groups of cells allowing for the conduction of water from the plants base to its upper reaches, working against the earth's force of gravity. This process, known as transpiration, is absent in the more primitive, non-vascular land plants, like mosses and liverworts.

Non-vascular land plants are those that, although they can live on land, are not capable of conducting water through any amount of plant tissue. They have to live in permanently moist areas and cannot tolerate dry places. Non-vascular plants include the mosses, liverworts, and the hornworts—all of which the fossil record is poor and incomplete. Mosses and liverworts, being more primitive than vascular plants, are presumed to have existed and lived in damp locations of the earth's land areas prior to the existence of vascular plants. Vascular plants first appear in the Silurian Period, essentially evolving to cover large portions of the earth's land masses as both large trees and smaller, herbaceous plants.

Today, leafless plants exist in a variety of types. Most of these types are not primitive, but rather represent former leaf bearing plants that underwent a type of "evolutionary degeneracy." These plants have reverted back to a form resembling primitive leafless plants. Examples of these are wild onions, grass, and even a plant that not only lost its leaves but also its chloroplasts and therefore its ability to photosynthesize. This is a flowering plant known as a dodder. All of the above mentioned plants are angiosperms, advanced flower producing plants that reproduce by the use of seeds. Psilophytes, on the other hand, lack leaves because they had yet to evolve them—leaves being a more advanced plant condition, an adaptation to increase the plant's photosynthetic surface-area. Leaves enable the plant to be more efficient in the process of photosynthesis. With leaves and the increased surface area made by them, the plant is able to catch more "rays." In the early Paleozoic, this improvement had not yet evolved, **plants of the time being truly primitive**. Only limited populations of psilophytes live today, primarily *psilotum nudum* of New Zealand, a **living fossil**.

Liverworts: like mosses, liverworts are non-vascular, primitive land plants and, as with mosses, they also have a sparse fossil record of which the best examples have been found in the Rhynie Chert of Scotland. Both liverworts and mosses cannot tolerate direct exposure to the sun. Because of this, they live in protected, shady, and moist environments—a requirement which probably limited their distribution in the geologic past as well. Both liverworts and mosses are rare fossils.

Another group of modern liverworts. These examples are clumped together in their moist habitat.

Dodder: a leafless plant that is "degenerate" from a more complex plant, not a primitive one. A dodder is an angiosperm, a flowering plant that has lost not only its leaves, but also its ability to photosynthesize. As a consequence, it has to parasitize other plants for its physiological energy. Most leafless plants seen today are like this, plants that have lost some fundamental feature. Often leaves are that fundamental feature lost, and with that the ability to photosynthesize, as in the dodder's case.

Soil, Regolith, and Paleosoils (Paleosols)

With the subject of land plants comes the matter of soil! **Soil is a loose mixture of plant material (humus) and inorganic (rock) material (known as regolith). To be true soil, both of these two components have to be present. Prior to the existence of land plants, there could not have been any soil on the earth's surface. Also, if there is no life on a planetary surface, like that of the Moon, there can be no soil on that planet.** There is a record of loose surface material existing on the earth's surface that goes deep into the geologic past. Such material is often associated with what are known as unconformities. These can contain carbonaceous material, grey or black material that may have originated from land plants; if it did, it is known as a paleosoil (or paleosol). Some paleosoils contain actual fossil roots . The late Paleozoic plant genus *Stigmaria* is an example that can occur in paleosoils formed as a substrate underneath the coal forests. Today, stigmaria occurs in what are known as underclays in coal geology, clay that underlies the coal of a coal seam.

Hiatus involving two sequences of strata separated by a paleosoil. Known geologically as an unconformity, older Lower Ordovician rocks (right) have been covered by Middle Ordovician sandstone (St. Peter Sandstone) with the two sequences separated by a zone of regolith (or paleosoil(?)). Underneath the man's hammer is an ancient (~460 million year old) regolith zone. The light colored rock at the right (Cotter Dolomite) makes up an ancient knoll carved from this rock by erosion a few million years after the dolomite itself was deposited in shallow seas some 470 million years ago. A zone of regolith (or soil(?)) overlies, or covers, this knoll, which in turn is covered by the 450 million year old St. Peter Sandstone on the left. Like Sherlock Holmes's questioning the importance of the barking dog, ***the question here is in the absence of humus, were there any land plants living on the surface of this regolith (or fossil soil(?))? If so, what were they?*** No fossils (plant or otherwise) are present in this ancient regolith.

The same outcrop on I-44 labeled to show what is going-on geologically.

Late Precambrian fossil regolith: a major erosional unconformity (and its associated paleosoil) exists over a large portion of North America where fossil evidence of non-vascular plants like mosses or liverworts should (or might) be found ... but unfortunately have not! Shown here, in eastern Ontario, is an unconformity between Paleozoic (Ordovician) limestone and Precambrian (Proterozoic) quartzite. The hubcap is resting on Pre-Ordovician regolith (or paleosoil(?)).

Beartooth Butte, NW Wyoming. The reddish layer near the top of this butte (or mountain) is a Devonian stream deposit. It was deposited in either fresh or brackish water during the earliest part of the Devonian Period, some 390 million years ago. The sediment-filled stream channel is best developed where the red layers are thickest to the right of center. This river channel contains fossils of primitive, leafless plants known as psilophytes—the oldest, undoubted land plants. Lower potions of the butte are composed of Upper Ordovician limestone and the top of Mississippian age limestone (Madison Limestone).

Closer view of Beartooth Butte with relevant strata labeled. The Devonian stream channel is again to the right of center.

Early Paleozoic Land-plant-like Fossils

Land plants, unlike those that live in water, need strong support to deal with the force of gravity, as well as with harsher phenomena like wind, drought, and direct sunlight—all phenomena associated with living on land. Other difficulties to be surmounted are reproduction and desiccation prevention, but the main purpose of a stem or trunk is to deal with that gravitational force. This was accomplished by developing an anchoring root system and a framework composed of a tough polysaccharide (cellulose), a substance that also leaves a distinct carbonaceous residue, preserving well in fossils. What would be considered fragments of land plants if found in younger strata are sometimes found locally in rocks of Cambrian, Ordovician, and Silurian age.

Undoubted fossil psilophytes first appear in the Late Silurian, and these had spread over all land areas by the Early Devonian. By the Mid-Devonian, they had to compete with the true ferns, the lycopods, and with *Archaeopteris*. *Archaeopteris* is a plant some paleobotanists consider to be an early form of seed fern and one of the earliest land plants to leave attractive fossils. Psilophytes, by contrast, were (and are) small, herbaceous plants that, because of their lack of leaves and indistinct roots, are rather plain looking—their fossils not usually eliciting much excitement with unsophisticated fossil aficionados either. Psilophytes should be more appreciated than they are, as they represent the first spread of true land plants over the planet, psilophytes being the first vascular plants. Land plants capable of conducting water from the soil into the plant itself is a feat which mosses and liverworts are unable to do. Mosses and liverworts are often cited as the earliest land plants; however, without the vascular tissue of psilophytes, they were very limited as to the environment in which they could live. To survive, mosses and liverworts have to have a shaded environment in perpetually damp places, such as underneath bluffs or in deep depressions where direct sunlight is absent. Also, because of the limited environment in which non-vascular land plants can live, their fossil record is poor and sketchy. The fossil record of the psilophytes is much better as they could tolerate a greater range of environments and thus live over a larger portion of the earth's land areas. The introduction of psilophytes also ushered in the last of the earth's "stages," the Green-Earth-Stage, which ended the "White-Earth-Stage," a time when little or no life occupied the planet's land areas.

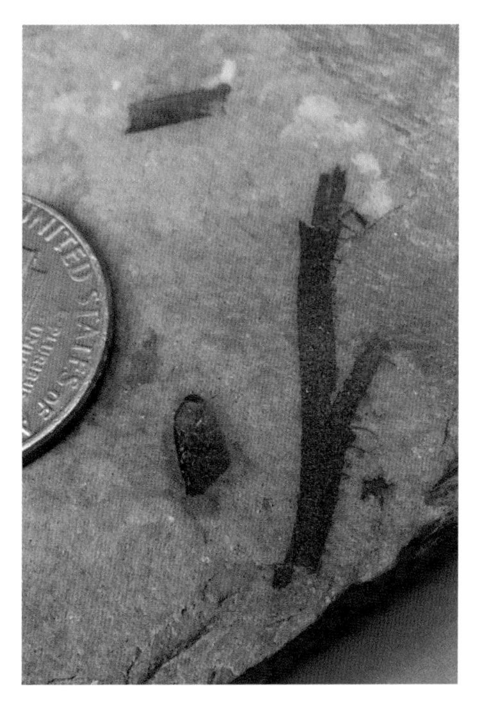

Psilophyte: a portion of a vascular plant stem from Beartooth Butte, Wyoming. It has branches coming off of the main stem but no leaves—land plants with leaves had yet to appear in the Early Devonian. These psilophyte fragments were carried by the same ancient stream that deposited the red channel seen in the photos of Beartooth Butte.

Psilophyte stems: note how the stem at the left has been twisted. The carbonaceous material that forms these compression fossils represents a residue of the original plant. Vascular plants, to form a firm structure, utilize cellulose, a rigid polysaccharide that enables the plant to stand-up against gravity. Marine plants usually lack this compound, which is widely associated with land plants and (of course) is the major component of wood.

Another branching psilophyte stem fragment. Lower Devonian, Beartooth Butte, Wyoming.

Psilophyte(?) fragment: a carbonaceous compression fossil. The checkered pattern is a carbonaceous residue, a material especially characteristic of coalified cellulose. If this were found in younger strata there would be no question as to its being a portion of a fossil land plant, possibly a psilophyte. This plant, however, is from strata of lowermost Upper Cambrian age, a time when land plants are not recognized to have existed (although spores believed to be from land plants have been reported from the Cambrian). The fact is **that the point in geologic time when land plants first appeared on the earth is still unclear and "muddy."** These fossils came from excavations for a Wal-Mart store near Fredericktown, Missouri.

Cambrian land plants(?): carbonaceous compressions like this occur sporadically in Lower Paleozoic strata, especially associated with shale beds. If these fossils were found in younger strata, they would be considered as representative of land plants. They are preserved as compressions, a carbonaceous film of plant material compressed in what originally was clay, and now is shale. Such carbonaceous (coal-like) material generally is considered to have been derived from cellulose and cellulose is characteristic of, but not totally restricted to, land plants. Upper Cambrian (false Davis) shale within the Bonneterre Formation, Fredericktown, Missouri (from an excavation for a Wal-Mart store.)

More (questionable) Wal-Mart psilophytes.

Exposure of the Upper Devonian Grassy Creek Shale below overhanging Lower Mississippian age limestone, Highway 21 north of Hillsboro, Missouri. The plant-fossil zone is one of **leafless plants** that have been considered to be **psilophytes**. Reproductive structures that might determine these fossils to be psilophytes have not been observed and these fossils might be compressions of some type of green algae. They apparently lived near the edge of the Ozark Uplift adjacent to shallow seas.

Psilophytes: compression fossils of what may be a primitive leafless vascular land plant. Upper Devonian, Grassy Creek Formation, Hillsboro, Missouri. Psilophytes are considered to be the earliest vascular land plants. Vascular tissue allowed water to be carried upward against the pull of gravity through the process known as transpiration. Non-vascular plants like liverworts and mosses have to live in sheltered environments to prevent desiccation. Vascular land plants avoid desiccation by conducting water from the plant's substrate (roots) through elongate connected cells known as vascular bundles. These leafless plants have left a carbonaceous film suggestive of cellulose and **cellulose is especially characteristic of land plants.** These fossil plants (somehow) suggest a form of algae, however, rather than true early land plants—psilophytes.

Leafless plant compression in siltstone exposed in the making of the cut from Grassy Creek Shale, Hillsboro, Missouri. (Value range G.)

Leafless plant(s) preserved as carbonaceous compressions in tan shale. Grassy Creek Formation, Upper Devonian, Jefferson County, Missouri. Nineteenth century paleobotanist J. William Dawson, commenting on plant fossils like this state's, "*Psilophyton*, in every country is one of the most characteristic plants of the Devonian period thought, when imperfectly preserved, is often relegated by careless and unskilled observers to the all-engulfing group of fucoids." (Fucoids are brown algae, large seaweeds). These fossil plants are still questionable about what they really represent.

Sawdonia ornata: this Lower Devonian plant was described by Canadian geologist-paleontologist J. William Dawson, who believed it to be a type of primitive leafless plant. These occur in Lower Devonian rocks of the Gaspe region of Quebec, as well as in New Brunswick and northern Maine. **They were some of the first fossil psilophytes to be described.** They are preserved as compressions, the cellulose of the original plant having become a black, coaly residue, more distinct than the usually vague outline of so-called seaweeds. Cap-aux-Os, Gaspe region, Quebec. (Value range F.)

Sawdonia ornata: this psilophyte has small protuberances or spines on the stems. Lower Devonian, Cap-aux-Os, Quebec.

Sawdonia ornata: a complex mass of the stems of this leafless plant.

Psilophyton sp.: a larger form of this psilophyte from Cap-aux-Os, Quebec.

Archaeopteris

Archaeopteris is the earliest known land plant of any size. It has a distinctive fern-like type of foliage.

Archaeopteris sp.: is the earliest large land plant. Known from the Devonian Period, the fern-like foliage of *Archeopteris* is believed to have been borne on large, pithy trunks, which would have made them tree ferns. Casts of these pithy trunks are known as *Eospermatopteris*. sp. (Value range F.)

Archaeopteris frond: a typical frond of this early, fern-like land plant. Middle Devonian, Northern, New York.

Archaeopteris sp.

Archaeopteris sp.

Glossary

Carbonaceous material: black or grey carbon bearing material found in sediments and sedimentary rock. Carbonaceous material can be derived from humus that might indicate the presence of land plants, especially if the dark colored material is associated with a paleosoil.

Deep Geologic Time: that portion of geologic time that represents the time before the fossil record becomes obvious. This usually is considered to be what geologists refer to as the **Precambrian**.

Humus: the organic component of soil. Humus is altered cellulosic (cellulose containing) material derived from land plants. From what can be determined from the geologic record, the early earth appears to have lacked soil, as there were no land plants.

Paleosoil (or paleosol): ancient soil from the geologic past (now often rock). The oldest undoubted paleosoil is Devonian in age; older ones are questioned as to their having really been soil.

Regolith: loose, inorganic material occurring on a planetary surface. A type of regolith is found on the Moon or Mercury as well as on most asteroids. Mars also has regolith rather than soil, as it does not (as far as can be determined as of 2013) appear to have or to ever have had life.

Rhynie Chert: this is a rock (chert) that formed by silicification of part of a Devonian peat bog in Scotland. The Rhynie Chert yields structurally preserved plants, most showing cell structure. Both mosses and liverworts have been identified in the Rhynie Chert, the earliest clear fossil record of these non-vascular land plants.

Soil: regolith that contains humus. The only planet in the Solar System to have soil is (probably) the earth as humus is a product of the biosphere. (Mars might have some form of primitive life but probably does not). Loose material (regolith) on the Moon has sometimes been erroneously called soil (as during the first Apollo Moon Mission). The Moon has no life, therefore it cannot have soil, which contains humus and is a product of plant life.

Unconformity: a hiatus (or major break in a local geologic record) often occurring at the bottom of a sequence of rock strata. This may include sediment that at one time was soil or regolith. Unconformities formed during the last three hundred million years can contain carbonaceous material, which almost certainly was derived from plants, probably land plants. Older unconformities generally lack such material, suggesting that land plants probably were absent.

Archaeopteris halliana: Canadian postage stamp with frond of *Archaeopteris*. This Devonian plant is found not only in New York and Maine, but also in Quebec, New Brunswick, Nova Scotia, and Newfoundland.

Chapter Four
True and Seed Ferns of the Late Paleozoic
Ferns (Pteridophytes)

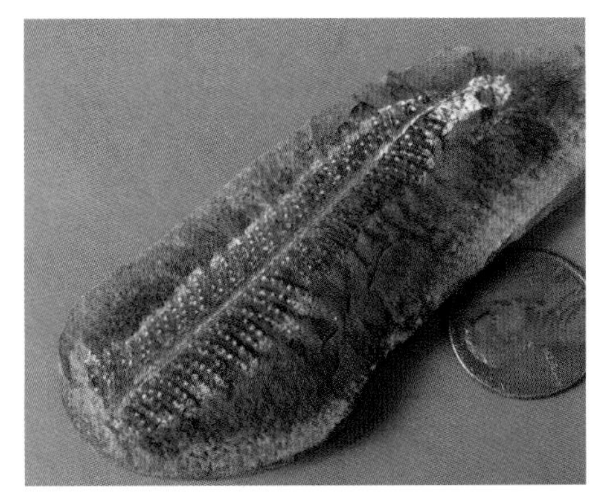

Pecopteris unita: fertile pinnae of a true fern. The spots on the pinnules are mineralized sori—sori being clusters of microspores borne on the ferns pinnules. Sori are characteristic of true ferns, which reproduce from spores. These spores can be released from the pinnules in great clouds where they scatter and fertilize the megaspores. This specimen is preserved in an ironstone concretion or nodule. Middle Pennsylvanian, Terre Haute, Indiana.

Two types of fern-like vegetation dominate late Paleozoic floras. Some of these fossils are from **true ferns**, members of the plant Division Pteridophyta. Others are from what are known as **seed ferns**, an extinct Division (Pteridospermophyta) of fern-like plants that reproduced through seeds. Representatives of both of these divisions are found associated with late Paleozoic (especially Pennsylvanian) coal swamp floras; however, it is usually difficult to determine which group specific fern-like foliage belongs as the diagnostic reproductive structures are not usually preserved.

Large portions of the U.S. Midwest are underlain by late Paleozoic strata. This occurs primarily in Illinois, northern Missouri, Indiana, Iowa, and in the eastern parts of Nebraska, Kansas, and Oklahoma. Usually these rocks consist of shale, thin limestone beds, and sandstone, rocks that at times are associated with coal beds. These coal beds occur in a somewhat regular, repeating order known as cyclothems.

A cyclothem consists of a limestone or marine shale bed (the marine portion) followed by an overlying coal bed and (non-marine) shale or sandstone layer, a sequence often found repeating over and over. The non-marine portion of the cyclothem usually formed in swamps—coal swamps were the homes of most of the ferns of this chapter. These swamps were the site of accumulation of major amounts of plant material that became the material responsible for coal beds. The marine portion of a cyclothem represents a time of higher sea level, which is believed to have been caused by melting of glaciers located in the southern hemisphere. Localized areas of these rocks, especially shales (which in the continental interior can be more like compacted clay), have preserved the vegetation of the "coal swamps." This preserved vegetation includes the leaves of fern-like plants illustrated in this chapter. Although covering a large area, this flora had rather limited diversity.

A "fiddlehead" or fern crozier: an early growth stage characteristic of a true fern. True ferns, when sprouting, take on this fiddle-head shape. From Mid-Pennsylvanian shale, Windsor, Henry County, Missouri. (Value range F.)

Both true ferns and seed ferns were the dominant plants of the coal swamps. These two categories of plants are placed together here not only because they are often found together, but also because, unless reproductive bodies are found associated with their foliage, it is impossible to separate one group from the other. With seed ferns especially, rarely are seeds actually found on the fossil plant, rather they are found separately, usually where they became concentrated while being carried by streams. More commonly found are the fertile portions of fern foliage bearing sori, clumps of spores borne on the underside of a fern leaf or frond. These more common fossils are representative of the true ferns.

Dolorotheca sp.: part of a *Dolorotheca* bell, a structure that bore the megaspores of tree ferns, which really were true ferns. Microspores, found on the leaves of ferns (sori), fertilize the megaspores in the bell (which is also known as a synagium). The fertile megaspores can then be broadcast from the bell to form potential new fern plants.

Permian forest: the last period of the Paleozoic Era—the Permian saw many plants go extinct that were so widespread in the latter part of the Paleozoic Era. This illustration from an early twentieth century Russian publication shows chordates at the left, arthrophytes on the right, and herbaceous ferns at the bottom. Permian conditions were generally drier than were those of the preceding Pennsylvanian (Upper Carboniferous).

Late Carboniferous (Pennsylvanian) tree ferns. Reconstruction of extinct plants from portions found as fossils can be tricky. This reconstruction from the late nineteenth century is surprisingly close to what is interpreted today. Interpretations vary, however, with different artists' viewpoints differing as to how these ancient plants actually grew. Tree ferns often dominate Late Carboniferous reconstructions and dioramas. (From J. William Dawson's *The Earth and Man.*)

Trigonocarpon sp.: a single, internal mold of a seed-fern seed. Lower Pennsylvanian, southern Illinois. (Value range F.)

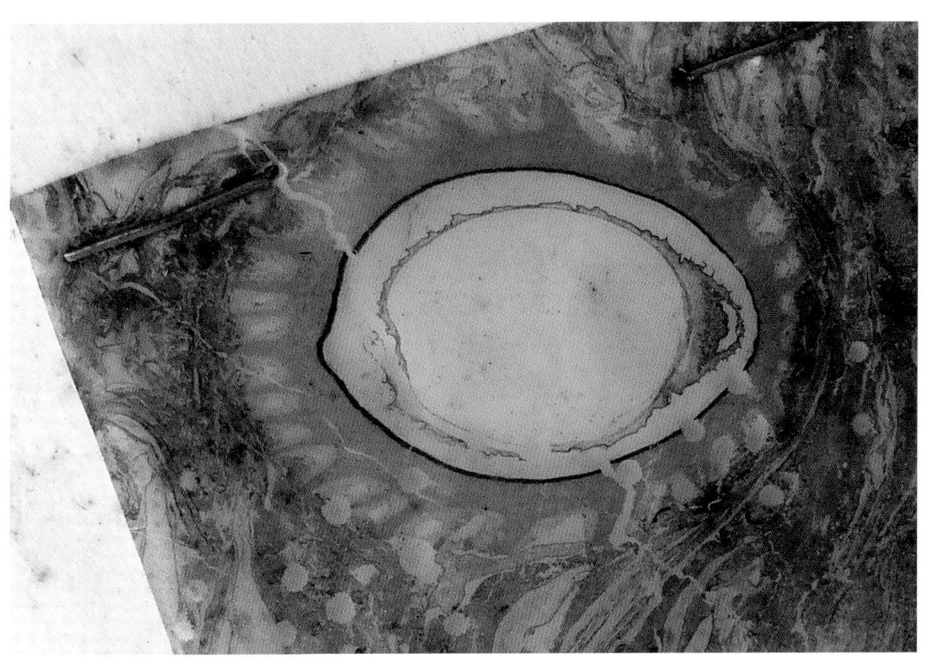

Seed fern or pteridosperm seed from a coal ball peel. Coal balls are formed when some of the plant material, which otherwise would become coal, becomes permeated with mineral material, primarily calcium carbonate. This mineral infiltration can then preserve plant tissue with great detail and fidelity, which includes this seed's structure. Seed ferns represent an extinct Paleozoic plant Division—the equivalent of a phylum in animals. Middle Pennsylvanian, Berryville, Illinois.

Seeds of seed ferns: true ferns (the ferns living today) do not reproduce by seeds. Seed ferns represent an evolutionary experiment in the development of the seed—**an experiment that ended at the close of the Paleozoic Era when seed ferns went extinct.** The seeds of pteridosperms (or seed ferns) are sometimes found in late Paleozoic plant associations. These come from a fern-like group of plants—an extinct taxonomic group of plants (Division)—equalivent to a Phylum in the Animal Kingdom. As such, the plants that bore them were not true ferns, rather they were fern-like in vegetation. This vegetation evolved **independently** from the **true ferns. True ferns, unlike seed ferns,** are still living today. Seed ferns, by contrast, went extinct at the Permian extinction event at the end of the Paleozoic Era.

Initial excavation for I-70 in 1954 in northern St. Louis County, one of the first interstate highways. Fern foliage fossils at the time were exposed at the bottom of the valley seen here. The bed was partially covered when the highway was completed; later urbanization covered the rest of the fern bearing outcrops. The boulders in the foreground were pushed by dozers to this position. They came from limestone beds underlying the fern bearing beds and are of marine origin. The fern bearing bed is in non-marine sediments.

Taxonomy of Upper Carboniferous (Pennsylvanian) Coal Swamp Plants

A considerable amount of literature exists on Pennsylvanian fern-like plants. This literature is found describing numerous species erected for various genera of fern foliage, as well as Linnaean names given to the various stems, parts of stems, and the larger trunks. Some of this taxonomy is a consequence of preservation, with various portions of a plant—such as foliage and wood—being preserved in different ways and thus given different names. The naming of different portions of plants is a reoccurring problem with plant fossils as different plant parts are almost always found separated and it is often unclear as to what goes with what. Some of this taxonomic proliferation also arises from the fact that Upper Carboniferous coal swamp floras occur over such a broad geographic area. As a consequence of this, different paleobotanists in various countries, states and provinces have differentiated species, often based upon preservation, trying to split the floras found over a large region into smaller entities. Often these geographically separated fossil floras are really morphologically identical. Regional occurrence of **modern plant floras** has also influenced thinking, so that the localized occurrences of modern plants becomes a template that is then applied to fossil floras. As a consequence, numerous taxonomic entities have been described that, although seemingly morphologically identical, are described as separate genera and species. It is the author's opinion that this broad geographic range is a major reason for the taxonomic richness in the literature of late Paleozoic coal swamp floras and that what actually exists is a more limited flora with not much variation, a cosmopolitan flora extending over a huge area of both eastern North America and Europe. What really is significant is how similar the floras of this time are and how big an area they covered.

Boulders of Mid-Pennsylvanian age occur just below the fern bearing horizon exposed in 2006 in northern St. Louis County. Unfortunately, the yellow shale at the locality contained no plant fossils.

Sandstone river-channel deposits (Warrensburg-Moberly Sandstone) overlain by shale beds. Tree fern trunks and fossil wood of tree ferns came from river-channel deposits in this excavation; however, foliage fossils were not present. Occurrence of fossil foliage is usually a localized thing. The top layer shown is the pavement of I-70, which was in the process of being rerouted. The vertical cylinders (which resemble erect fossil trees) are the concrete supports of former road signs.

Localized Occurrence of Outcrops of Pennsylvanian Fossil Plants, Especially Foliage

Fossil plants, especially foliage, can occur in distinct horizons that may come and go in the rock record of a particular region. Often preserved in what are referred to as shale lenses, plant fossil zones may show up in stream outcrops, in cuttings for roads, and in the shale beds overlying coal seams—coal that may be mined, exposing overlying fern bearing layers. A (locally) prolific fossil plant horizon occurs in the U.S. Midwest at the very top of Middle Pennsylvanian strata. Many of the "ferns" illustrated on the following pages came from this horizon, which, as mentioned above, is representative of a very cosmopolitan flora occurring over both Europe and North America.

Collecting fern leaf impressions from siltstone beds that underlie the Warrensburg-Moberly Sandstone in north St. Louis County, 1954.

Warrensburg-Moberly Fern-like Plants, St. Louis County, Missouri

A series of stream deposits of Mid-Pennsylvanian age cover sizeable portions of the U.S. Midwest. Locally shale and sandstone beds of these stream deposits can be rich in fossil vegetation. Shale beds contain fern-like foliage and sandstone beds contain the more robust portions of plants, like stems and trunks. Plant bearing beds in what is known as the Warrensburg-Moberly Formation were exposed and accessible from the 1930s to the 1980s in the St. Louis area. Afterwards most of these formations became covered by urbanization. The author collected these fossil plants from what were sometimes transitory outcrops starting in the early 1950s, some specimens of which are shown in the following images. To the best of the author's knowledge, the illustrations shown here are the only documentation of this flora from the St. Louis area.

These plant bearing beds make up part of what is known as the St. Louis Pennsylvanian outlier, a remnant of what, at one time, was strata connected to the Illinois Basin, which lies to the east. Interestingly, this area of downwardly deflected strata is a consequence of the same geologic forces that later deflected the Mississippi River to the east before the Missouri River runs into it, producing the geography instrumental in the founding and location of the city of St. Louis itself. (Geology often explains aspects of "the lay of the land," which in turn influenced settlement). The band of Pennsylvanian strata associated with this eastward bend of the Mississippi is also the source of the fossils shown here.

Close-up of fern-leaf bearing siltstone beds.

Fern bearing siltstone beds in the Pleasanton Group, late Middle Pennsylvanian, St. Louis County, 1954.

Pecopteris sp.: vague fern frond impression in sandstone from the previously shown outcrop. Coarse sediments, like this iron-rich sandstone, preserve plant foliage, but unlike fine sediments like clay or silt, preservation in sandstone is poor. Robust portions of plants like stems, trunks, or roots generally are found in coarser sediments like sandstone. The pinnules and pinna of this frond are barely visible. Warrensburg-Moberly Sandstone. Upper Middle Pennsylvanian, St. Louis County, Missouri.

Massive sandstone boulder that contained fern frond impressions. Such fronds are uncommon in sandstone. It normally doesn't preserve them or preserves they very poorly, as is the case here.

Pecopteris sp. (left) and *Neuropteris flexuosa* fronds (right): *Pecopteris* is a true fern. *Neuropteris* is believed to be a seed fern. Shale of the upper Mid-Pennsylvanian Pleasanton Group, northern St. Louis County, Missouri.

Pecopteris cf. *P. squamosa* (left) and *Neuropteris flexuosa* (right.)

Fern-bearing yellow siltstone overlain by massive Warrensburg-Moberly Sandstone from which the following plants were collected, St. Louis County, 1955.

Pecopteris cf. *P. squamosa* Lesquereux. *Pecopteris* (also known as *Asterotheca*) is a common Pennsylvanian fern. It is preserved here as a compression in iron-rich, oxidized shale of the Pleasanton Group, St. Louis County, Missouri.

Pecopteris cf. *P. squamosa* (left): a terminal pinnae (ultimate pinnae) is on the left, penultimate pinnae are on the right. These came from a cluster which was possibly from the same frond.

Neuropteris flexuosa Lesquereux. Another specimen of this fern from the previously shown outcrop.

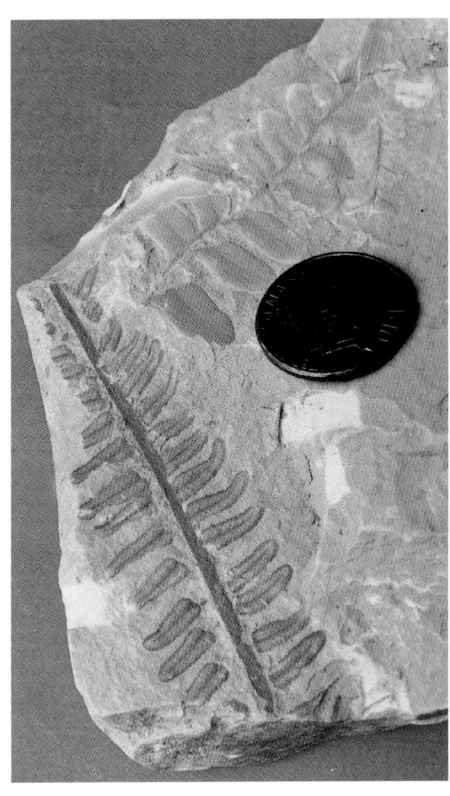

Pecopteris sp. (left): pinnules have separated from the fern's stem as a consequence of compaction of the original clay. *Neuropteris flexuosa* (top.)

Neuropteris vermicularis Lesquereux: this species of *Neuropteris* is especially characteristic of the latest Mid-Pennsylvanian, the age of the St. Louis County basin Pleasanton fossil plants. This species is found in earlier Pennsylvanian plant horizons like the Mazon Creek flora. Here the specimens are larger. (A partial specimen of the arthrophyte *Annularia* can be seen to the left.) (Value range G.)

Pecopteris sp. (part and counterpart): small *Pecopteris* pinnae like this can be common in Pennsylvanian shales when they contain some evidence of the presence of fossil plants.

Neuropteris flexuosa: here the pinnules have been preserved as a film of iron oxide, a characteristic of fossil plants preserved in oxidized sediments like these and often characteristic of shales of the late middle Pennsylvanian Pleasanton Group.

Neuropteris vermicularis: this appears distinctive from previously illustrated specimens of this species as it apparently was somewhat waterlogged prior to being buried in a clay slurry. This resulted in its being more elongate and "wormlike," the source of the species name.

Neuropteris flexuosa frond: the first frond found in the St. Louis County outcrops.

Group of *Pecopteris* pinnae from different portions of a frond

Warrensburg, Missouri, Compression Fossils in Grey Shale

Fossil plants from near Warrensburg, Missouri (Knob Noster site of Chapter 9), of the same age and preserved under similar conditions as those of the St. Louis area, have been extensively distributed through some fossil dealers. The author has seen numerous specimens from the Warrensburg, Missouri, area show up at MAPS EXPO, a yearly fossil show in Macomb, Illinois. The specimens are preserved either in yellowish shale similar to that of the St. Louis area or are preserved as black carbonaceous compressions in grey shale. Of the latter, the carbonized plant material often flakes off and specimens are best coated with a fixative, such a fixative may, however, create a gloss that can interfere with photography.

Neuropteris flexuosa frond: this is a characteristic "fern" from grey shales of the Pleasanton Group in various parts of the U.S. Midwest. A large number of these ferns, collected from creek beds in the Knob Noster area near Warrensburg, Missouri, have circulated among the collecting community from 2000 to 2004 through the MAPS Expo fossil shows.

Neuropteris flexuosa frond: carbonaceous compression where the "carbon" has partially flaked away. (Value range G.)

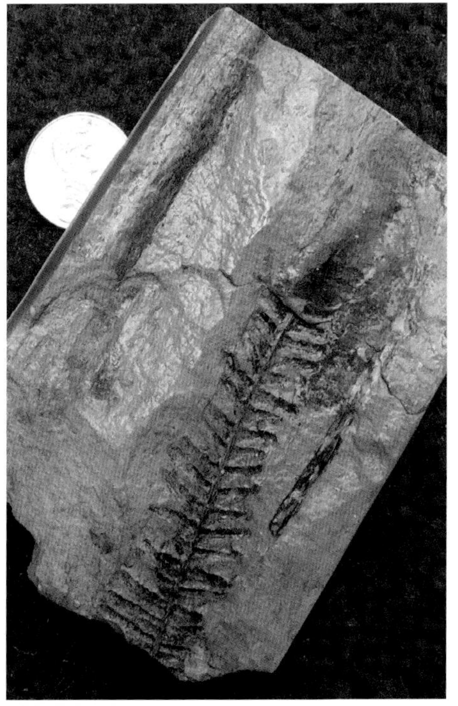

Pecopteris cf. *P. squamosa* penultimate pinnae.

Pecopteris squamosa ultimate pinnae.

Neuropteris flexuosa (partial compression). (Value range H.)

Pecopteris sp.: left, broad form; right, slender portion of pinna near termination of the frond.

Plant Compressions, Northeastern Missouri

Pennsylvanian strata cover a large portion of northern Missouri. In the northeastern part of the state and northward into Iowa this strata fills in and covers an irregular surface of older Mississippian age rocks. Depressions in this ancient surface of Mississippian limestone became filled in by Middle Pennsylvanian age sediments that (sometimes) can contain localized occurrences of fossil plants. Sometimes sediments in these depressions preserved rare upland plants of the Pennsylvania Period rather than the more common plants of the coal forests. Shown here is a flora from one of these localized areas in Shelby County, Missouri. It is of late Middle Pennsylvanian age.

Neuropteris flexuosa: a nice specimen in siltstone. (Value range F.)

Cyclopteris sp.: one of the few broad leaves found in coal swamp floras. *Cyclopteris* is considered to have come from the same plant as *Neuropteris Cyclopteris,* or (*Meganeuropteris*) *scheuchzeri* is believed to have been borne at the base of a frond. (Value range G.)

Neuropteris sp. (small fern). (Value range F.)

Frond of *Neuropteris flexuosa*: complete fronds are difficult to collect from natural outcrops, requiring as they do the extraction of relatively large slabs of shale. They are more readily seen and collected from rock exposed in man-made excavations, like coal mines. (Value range F.)

Lower Pennsylvanian Shale, Filling Sinkholes

Pennsylvanian sediments of Lower Pennsylvanian age can, like those of the middle Pennsylvanian, be found filling ancient sinkholes (also known as paleokarsts) developed in underlying limestone. These less commonly seen lower Pennsylvanian sediments are present in Missouri, Illinois, and Iowa. Locally, they can contain fossil fern leaves. Sometimes these ferns are genera **not found in younger** (Middle) Pennsylvanian strata associated with coal swamp floras. Some of these Lower Pennsylvanian plant genera are also believed to be representative of a more upland, drier environment than were the plants of the coal swamps.

Lower Pennsylvanian age strata in outcrop near the Ohio River, southern Indiana. Over a large portion of the U.S. Midwest a major period of erosion took place near the end of the Mississippian Period. This produces, over a sizeable area, an unconformity (hiatus in the rock record), sometimes characterized by the presence of a paleosoil. It is partially for this hiatus that the Mississippian and Pennsylvanian periods were distinguished from the European Carboniferous. Pennsylvanian strata can cover this irregular (unconformable) surface, filling in low places. These beds, of lowermost Pennsylvanian age, lie just above Mississippian limestone, which crops out just to the south in Kentucky. (Photo taken 1955.)

Neuropteris sp.: a puzzling variation of this common genus.

Partial frond of *Pecopteris* from shale layers exposed in a creek bed. The slab is stained with limonite, a characteristic of plant compressions collected from natural outcrops.

Water flowing over Lower Pennsylvanian strata similar to that above but in Kentucky. Few plant fossils are present in these thick shale and siltstones unless a coal bed is present.

Neuropteris sp.: tree fern foliage from Lower Pennsylvanian strata. Locally occurring shale beds can fill depressions (probably ancient sinkholes) in underlying Mississippian limestone of the western Ozarks of Missouri. In the U.S. midcontinent during the latter part of the Mississippian Period, a great deal of weathering and solution took place with the older Mississippian limestones. Very localized Pennsylvanian sediments later filled in these depressions, some of which contain fossil plants of what is an upland flora rather than the more normal coal swamp flora.

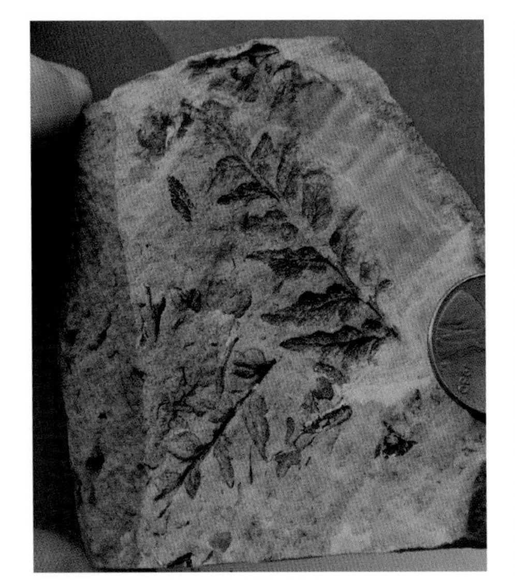

Mariopteris muricata: seed fern, Lower Pennsylvanian, Caseyville Formation, Wyoming Hill, Muscatine County, Iowa. Lower Pennsylvanian strata in Iowa, as in Missouri, can fill depressions (or paleokarsts) in underlying limestone. These paleokarst fillings can contain plant fossils of a less common type than those of the coal swamp flora. These less frequently seen floras are believed to represent a more upland, drier climate than fern genera like *Pecopteris*, which is characteristic of the coal swamps.

Mariopteris muricata: close-up of previously shown specimen.

Civil War Veteran John H. Britt's Fossil Plants from Western Missouri

Numerous well preserved plant fossils have come from various localities in Henry and Bates counties in Missouri, located at the northwest edge of the Ozark Uplift. Some of these, especially in the late nineteenth century, came from shale overlying coal beds formed in association with solution structures developed in underlying Mississippian limestone, which is an occurrence similar to the fossil plants of northeastern Missouri. These localized coal deposits were extensively mined in the late nineteenth and early twentieth centuries, where overlying shale layers yielded fabulously large and complete fossil plants. Some of these consisted of complete fern fronds, a few even with attached seeds showing that they were seed ferns, unique plants that are extinct today. Some of this flora was illustrated in USGS (United States Geological Survey) Monograph 37 by David White, a paleobotanist and coal geologist. The fossil plants illustrated in this work were collected extensively by John H. Britts, a Civil War veteran and physician who became a resident of nearby Clinton, Missouri. Dr. Britt's collections formed not only the basis for White's USGS monograph but also were distributed to many institutions. Many fine fossil plants found their way through Dr. Britts into the cabinets of colleges in the U.S. Midwest. Britts' specimens can be identified by the date collected and the initials J B neatly inscribed on the shale slabs. As some Midwestern colleges have recently divested themselves of their paleontological collections, some of Britts' specimens have recently showed up on the fossil market.

UNITED STATES GEOLOGICAL SURVEY
CHARLES D. WALCOTT, DIRECTOR

FOSSIL FLORA
OF THE
LOWER COAL MEASURES
OF
MISSOURI
BY
DAVID WHITE

WASHINGTON
GOVERNMENT PRINTING OFFICE
1899

One of the better documented Pennsylvanian floras of the Midwest occurs in southwest Missouri on the edge of the Ozark Uplift in Henry County. This flora, from the lower part of the Middle Pennsylvanian, was published in a USGS (United States Geological Survey) monograph in 1899. The flora discussed in Monograph 37 was described by David White, a paleobotanist and coal geologist. This flora was collected from shales that overlie a localized coal seam associated with sediments, which filled depressions developed earlier in underlying Mississippian limestone. These depressions were formed from a major period of erosion and solution in late Mississippian time, which also resulted in a major unconformity between Mississippian and Pennsylvanian strata in the U.S. Midwest. Pennsylvanian sediments filling these depressions can sometimes preserve fossil plants with superb fidelity. Many excellent fossil plants from these occurrences were collected by John Britts of Clinton, Missouri, including those illustrated in White's monograph. Many of Dr. Britts' fossil plants came from shales overlying coal seams exposed in small coal mines, many becoming parts of fossil collections in colleges of the Midwest. Specimens from these old collections sometimes show up on the fossil market, as many geology departments in the U.S. no longer appear interested in paleontology.

Pecopteris exculpis: a large frond collected by John Britts. Similar specimens, all of which were collected prior to 1899, were illustrated in White's Monograph. This specimen was not, as it was collected in 1904. (Value range D, of historic interest).

Neuropteris sp.: a frond of this widespread fern genus from the Henry County **sinkhole** coal mines, collected by John Britts (note the J B and 1904 date at the bottom). (Value range E.)

A rotary drill core sample from Middle Pennsylvanian shale containing a *Neuropteris* specimen, Henry County, Missouri. From a sample drilled for a guided missile silo—part of the cold war legacy.

The date when John Britts' specimens were collected is inscribed in the manner shown on this specimen, in this case November 19, 1904. Below the date is inscribed J B for John Britts of Henry County, Missouri.

Alethopteris serli: a specimen from the collection of the St. Louis Science Center. These specimens were given to the St. Louis Science Museum by the geology department of St. Louis University. They were probably obtained from J. Britts in the early part of the twentieth century. It was common for paleontological authors in the nineteenth and early twentieth centuries to donate specimens of fossils on which they had published research texts and articles to geology departments and museums. As many of these institutions have taken a direction away from paleontology, these old collections with their "classic" fossils, from "classic" localities, are being recycled and sometimes appear on the fossil market.

Croweburg Coal exposed in a strip mining operation near Windsor, Henry County, Missouri, 1962. Grey shale overlying the coal seam has been removed with plant (and ironstone concretion) bearing layers piled on top of the spoil piles where they were then accessible, as in the shale pile to the right.

Western Missouri's Younger Middle Pennsylvanian Fossil Plants

Later, in the 1930s through the 1960s, coal beds younger than those mined near Clinton, Missouri (whose flora was originally described by David White), were strip mined. Mined especially was the Croweburg Coal, which yielded excellent fossil plants as compressions in grey shale as well as occurring in fossil bearing iron stone concretions similar to those of Mazon Creek and Braidwood, Illinois.

Shale exposed in the high wall of a coal strip pit near Windsor, Henry County, Missouri, in 1962. These grey shale layers are barren of fossils, except for the layers at water level, which occur just above the mined coal bed.

Smaller *Neuropteris flexuosa* frond, Windsor, Missouri. (Value range E.)

Fern bearing shale slab from strata overlying the Croweburg Coal near Windsor, Henry County, Missouri. The Croweburg Coal, and the plant bearing horizon which occurs above it, is stratigraphically above those beds yielding the ferns collected by J. Britts. The Croweburg Coal and its correlatives underlie a large portion of the U.S. Midwest. (Value range E.)

Large frond of *Neuropteris flexuosa* from near Windsor, Missouri. These were collected in the 1960s from strip mining operations in the Windsor area. Reclamation of mining areas like this now prevents these nice fossil plants from being collected and thus salvaged for education and science. In the future, when coal mining is no longer done, we will lament that more of these fossil plants were not salvaged. (Value range D.)

Pecopteris grandis. Shale above the Croweburg Coal, Windsor, Henry County, Missouri.

Two fronds of foliage that are somewhat indeterminate as to genus, but are close to *Pecopteris*.

Pecopteris in shale above Croweburg Coal, Henry County, Missouri. Specimen in the collection of the St. Louis Science Center.

Group of *Aleothopteris:* the very distinct impressions of this genus of (possible) seed fern can be seen here to good effect. (Value range F.)

Southern Indiana and Kentucky Plant Fossils

Fossil plants almost identical to those from the younger strata of Henry County, Missouri, were (are) mined in southern Indiana and northern Kentucky. Both grey shales containing plant compressions and numerous ironstone nodules with plants (and more rarely animals) have come from strip mining operations near Terra Haute, Indiana. The fossils are also about the same age as the famous Mazon Creek ironstone concretions of northern Illinois.

Aleothopteris serli: the pinnules of *Aleothopteris* were thick and possibly waxy. They leave very distinct and sharp compressions in shale, as can be seen on this specimen. (Value range F.)

Pecopteris sp.: somewhat suggestive of *Aleothopteris*, some of the Pennsylvanian coal swamp fossil plants grade into each other rather than being distinct from each other, which is not the case with most plant genera and species (including ferns) living today. (*Sphenophyllum* to the right.) (Value range F.)

Neuropteris (Macroneuropteris) scheuchzeri "triad:" *Neuropteris scheuchzeri* generally is found as single leaves. Here is a group of three.

Odontopteris wortheni Lesquereux: an uncommon coal swamp fern with elongate, tongue-shaped leaves.

Sphenopteris sp. Brongniart: fern of a type found in both upland and coal swamp floras. (Value range E.)

Upland Pennsylvanian Plants and "Frilly" Ferns

Siltstone beds in Middle Pennsylvanian age strata sometimes yield ferns of a type different from those associated with the normal coal age flora of *Pecopteris, Neuropteris,* and *Aleothopteris*. These less common "frilly" ferns can be of the genera *Sphenopteris, Renaultia,* and *Diplothmena*. They are believed to represent a more "upland" flora, plants that came from a drier environment than those of the coal swamps.

A "man made hill" comprised of Pennsylvanian sediments (sandstone and siltstone) that once filled an ancient sinkhole (or other solution structure), which was developed in Mississippian limestone (St. Louis Limestone). The surrounding limestone was removed in quarry operations in the 1960s, leaving behind the filled sinkhole. There was a profound amount of weathering and removal of rock between the Mississippian and Pennsylvanian periods in the U.S. Midwest, enough time to form not only this filled sink but to remove many thick layers of older strata so that Pennsylvanian rocks could then lie upon Ordovican and even Cambrian strata. This hiatus is (or was) one of the reasons for establishment of the Mississippian and Pennsylvanian periods and distinguishing them from the European Carboniferous, which covers the same part of geologic time. Shale and refractory clay, deposited in these solution structures, can sometimes contain fossil plants different from those found in the coal swamp floras—more of an upland flora.

Renaultia gracilis Brongniart: note that the fern in the middle is brown rather than black. Brown compression material is more lignitic. The coalification process has not gone as far with this fern as it has with the other compressions on this slab. Something in the original plant material prevented coalification. Possibly the leaf was somewhat waxy, which hindered the coalification process, preserving complex organic compounds that are present in the more lignitic leaf.

Diplothema sp.: an aberrant fern that is more commonly associated with upland floras than with a coal swamp flora. Leaflets are deeply divided in this genus.

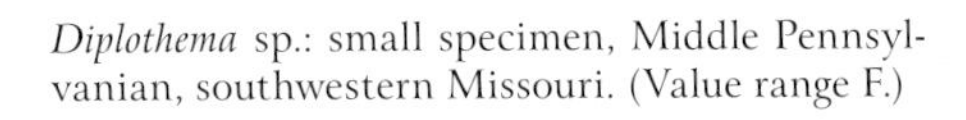

Diplothema sp.: small specimen, Middle Pennsylvanian, southwestern Missouri. (Value range F.)

Diplothema sp.: Middle Pennsylvanian, Dade County, Missouri.

Eremopteris sp.: single leaflet, Middle Pennsylvanian sinkhole deposits, Shelbina, Missouri. Another (probable), upland Pennsylvanian plant. (Value range G.)

Diplothema sp.: Middle Pennsylvanian, southwestern Missouri. (Value range F.)

Ozark Paleokarsts and Upland Floras

Pennsylvanian-age sediments in the Missouri Ozarks often occur in what are known as filled sinks, ancient sinkholes or depressions dissolved in Ordovician or even Cambrian dolomites and that were filled in with sediment during the Pennsylvanian Period. Sediments in these filled sinks, until recently, were mined extensively for refractory clay. Rarely, these "filled sink deposits" contained fossil plants. Unfortunately, the few that were found were not documented in the literature.

Pennsylvanian age sediments over the Ozarks can occur in filled sinkholes (so called paleokarsts). The sinkholes were formed prior to the Pennsylvanian by solution of Mississippian, Ordovician, and even Cambrian age limestones and dolomites. This is such a filled sink deposit that was prominently exposed along what at the time was Highway 66 at the northern part of Rolla, Missouri, later this became I-44. The Pennsylvanian sediments (white) are thin bedded layers. The sinkhole was filled in with Pennsylvanian sediments, but had developed in much older Lower Ordovician Dolomite (Cotter Formation), which is tan. What appears to be two large nodules (or a gigantic tongue) are rounded dolomite masses that formed part of the wall of the sediment-filled sinkhole. These ancient paleokarsts are (or were) the source of high quality refractory clay mined over parts of the northern Ozarks.

Neuropteris sp.: from refractory clay deposits six miles northeast of Rolla, Missouri.

Mazon Creek or Braidwood, Illinois, Fossil Bearing Ironstone Concretions

The Mazon Creek or Braidwood coal swamp flora (along with the marine Essex fauna) is considered to be one of the world's "paleontological windows." Originally discovered in outcrops along Mazon Creek (or Mazon River) in Grundy County, Illinois, strip mining of coal, especially in the 1930s and '40s, exposed large areas of fossil bearing shale, which overlie the coal bed in what are known as spoil piles. As these were washed by the rain and weathered, elliptical-shaped nodules or concretions eroded from the shale piles and the freeze-thaw cycle would then split them open along the major axis—split them usually where there was a fossil. These fossils have attracted numerous collectors, especially those from the Chicago area, who over the decades have collected hundreds of thousands of them. Such thorough collecting uncovered many scientifically valuable fossils of both plants and animals. The Field Museum in Chicago for decades actively assisted in this collecting, especially with the assistance of Eugene Richardson of that institution. The Field Museum has also published extensively on Mazon Creek fossils, as have some members of ESCONI—The Earth Science Club of Northern Illinois. These plant fossils are probably the best known and most widely distributed ones in North America.

Coal strip mine operation: four and more decades ago, shale beds overlying coal seams mined in the U.S. Midwest would be piled up by the shovel to produce "spoil piles." Shale layers which occurred just above the coal seam would remain at the top of the piles. This strategy would make available those layers bearing the fossil plants. Reclamation requirements now mandate that the piles be leveled and soil, removed before strip mining begins, be placed over the shale, thus burying it and also burying any fossils. Coal strip pits operating from the 1920s through the 1970s were a window of opportunity to obtain fossil plants from mining activities like this.

Spoil piles near Braidwood, Illinois, July 1953. These piles of shale, excavated during coal strip-mining in the 1930s and '40s, have been a major source of fine fossil plants preserved in ironstone concretions. The concretions weather from the shale and the freeze-thaw cycle splits them open, revealing (usually) a fossil plant. The Braidwood strip mines have produced hundreds of thousands of specimens, which were (and are) eagerly collected by locals, as well as by collectors from the Chicago area, which is about sixty miles to the northeast. Development and vegetation, however, have reduced the availability of easy collecting from this area during the past two decades.

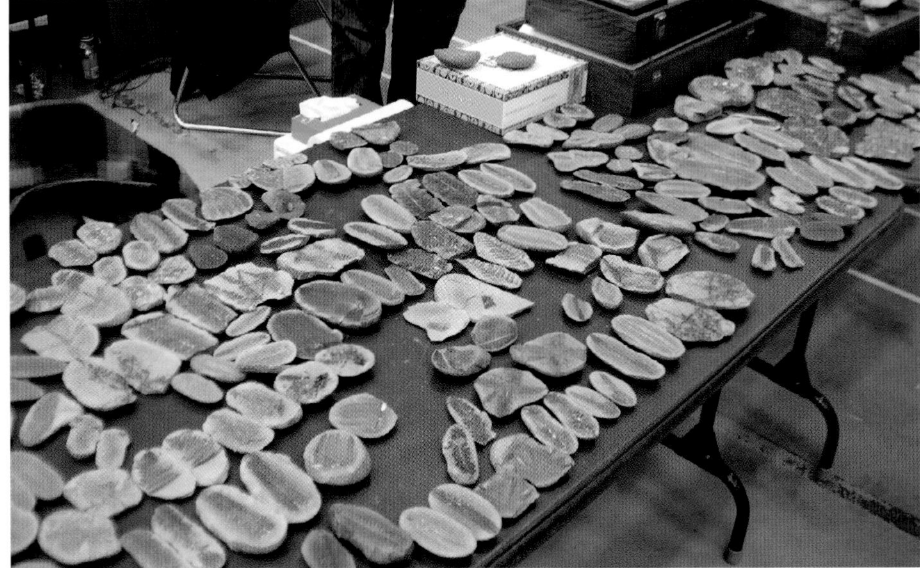

Plant bearing "Mazon Creek" concretions from the Braidwood-Coal City region at MAPS EXPO. Large numbers of these concretions with fossil plants (primarily *Pecopteris, Neuropteris,* and *Aleothopteris*) can often be found for sale or trade at rock and fossil fairs. Old collections continually supply specimens, as well as some still turning up from the old strip-pit spoil piles.

Leonard Stinchcomb, father of the author, at the Braidwood-Coal City spoil piles, July 1953.

"Mazon Creek" (Braidwood-Coal City area) plant bearing concretions in the collection of the St. Louis Science Center. Museums, college geology department, and private collections worldwide often contain numerous specimens of these attractive and interesting plant-bearing, ironstone concretions.

Pecopteris miltoni: probably the most common individual fern found in the "Mazon Creek" ironstone concretions.

Pecopteris miltoni: frond of what is probably the most common fossil fern found in the Braidwood spoil piles. (Value range F.)

Damaged frond of *Pecopteris miltoni*: this fern is also known as *Asterotheca*. *Asterotheca* and *Pecopteris* are identical, the genus *Asterotheca* being used when the author of a paleobotanical work considered that the illustrated fossil plant was a seed fern, *Pecopteris* being used when it was believed to be a true fern. (Value range F.)

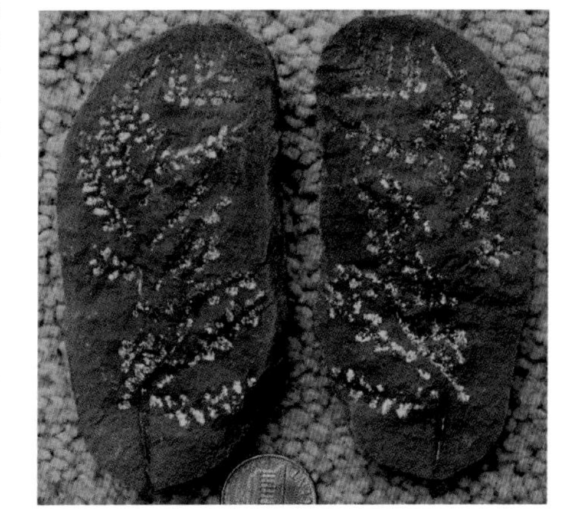

Fertile frond of *Pecopteris miltoni*: with a fern like this, spores compose much of the frond. (Value range F.)

Asterotheca (Pecopteris) crenulata. (Value range G.)

Asterotheca (Pecopteris) oreopteridia: a frond of a type still found occasionally in the Braidwood region. (Value range F.)

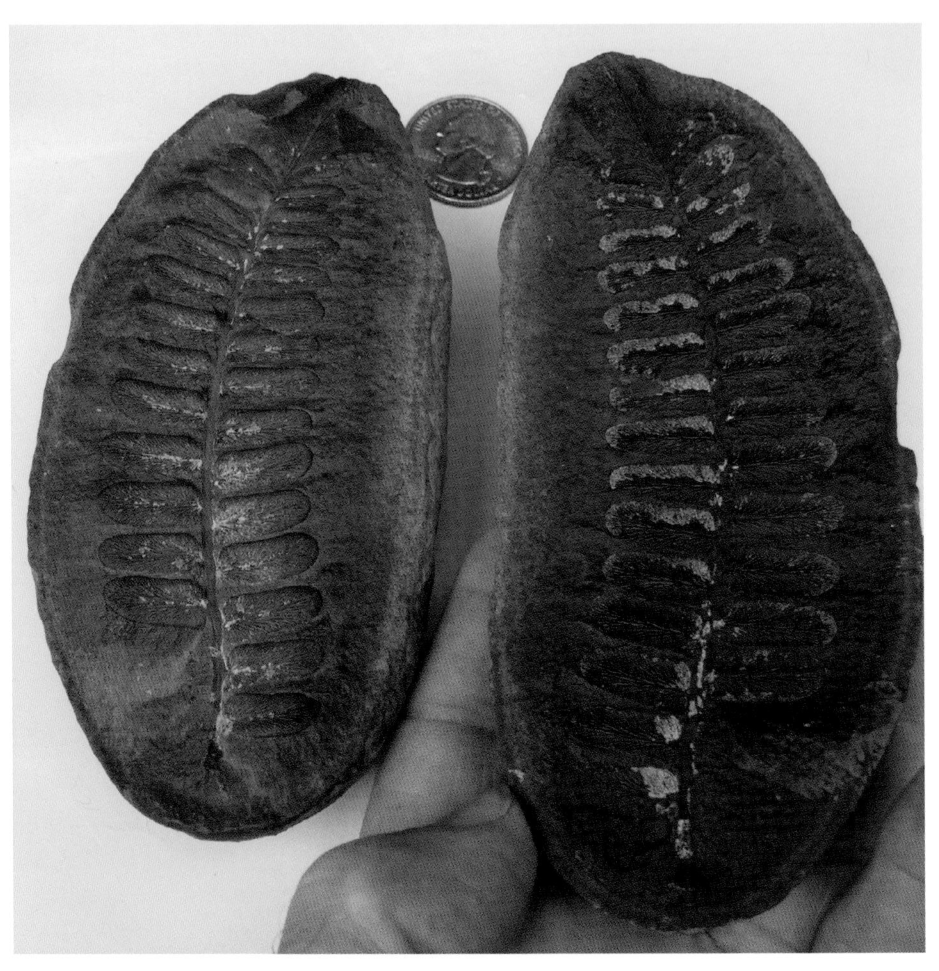

Neuropteris flexuosa: quality specimens from the Braidwood area like this are not seen too frequently today. This specimen was collected in 1953. Note that many Braidwood concretions and their ferns are large in size. Other similar occurrences in concretions (and there are a number of them) don't produce the large-sized ferns collected here.

Asterotheca (Pecopteris) sp.: specimen in unoxidized concretion. (Value range F.)

Small *Neuropteris* frond. (Value range G.)

Asterotheca (Pecopteris) sp.: single side of concretion.

Asterotheca (Pecopteris) crenulata: part and counterpart in an unoxidized ironstone concretion.

Neuropteris (Meganeuropteris) scheuchzeri (part and counterpart). (Value range F.)

Cyclopteris sp.: a more recently collected specimen of this globular shaped leaf. *Courtesy of John Stade.*

Odonthopteris sp.: a nice specimen showing part and counterpart of this less common "rounded leaf" genus. (Value range E.)

Cyclopteris sp.: some *Cyclopteris* leaves, especially if they have been somewhat water soaked and macerated, can resemble the wings of an insect. Insects are found associated with the Braidwood nodules, but they are much rarer than are the leaves. (Value range G.)

Cyclopteris sp.: this large, oval leaf is believed to be associated with *Neuropteris scheuchzeri*. It is quite different from other leaves of the coal swamp environment. Specimen collected in 1953. (Value range F.)

Neuropteris (Meganeuropteris) scheuchzeri in grey shale: ferns in grey shale from Braidwood are rare. They rarely were collected when they did occur. (Value range F.)

Neuropteris sp.: the leaf has been bent before being fossilized. (Value range F.)

Odonthopteris sp.: average specimen of this less commonly seen genus. Notice the rounded leaves, more rounded than *Neuropteris* of the previous photo. (Value range F.)

St. Clair, Pennsylvania *Alethopteris*

Strikingly beautiful "ferns," primarily of the genus **Aleothopteris**, have become widely distributed among fossil dealers and collectors. These come from black slate-like shale exposed in an anthracite coal mine near St. Clair, Pennsylvania. These fossils are striking as the large fronds are naturally outlined with white kaolin, which contrasts with the slate grey or black shale. The kaolin in some manner was introduced to the fossil plants during metamorphism of the coal when it was converted to anthracite.

Alethopteris serli: a large leaf in a large concretion. Large size is especially typical of the Braidwood fossil flora. Leaves of *Alethopteris* are especially large and often are well defined as the original leaf itself must have been thick and possibly waxy. (Value range F.)

Aleothopteris St. Clairi: these large *Aleothopteris* leaves are naturally outlined with white kaolin, which gives them a stunning contrast. They overlay anthracite coal beds in strata of Pennsylvanian age, which has been somewhat metamorphosed. The metamorphism introduced the kaolin, preserving them in a manner not found in other regions where the coal swamp flora is found. The St. Clair flora consists almost entirely of large *Aleothopteris* leaves. Examples of it have been widely distributed among collectors, dealers, and rock hounds. Slabs of these leaves make attractive decorative pieces as well. (Value range G.)

Slab with damaged pinnule. (Value range G.)

Complete pinnules. (Value range G.)

Compete pinnule. (Value range G.)

Typical St. Clair, Pennsylvania, slab. (Value range G.)

Group of ferns from slaty black shale from near St. Clair, Pennsylvania. These beautiful fossils were not completely covered at the site during mine reclamation. Numerous specimens (many collected after reclamation) have become effective tools in science education as they occur in large numbers in this previously mined area.

Other Eastern U.S. Late Paleozoic Plants

Excellent compression fossils of fern-like plants of Pennsylvanian age come from coal mines and outcrops of the Appalachian Mountains in Pennsylvania, West Virginia, Virginia, and eastern Ohio. Some of these have been widely distributed among collectors like the above mentioned *Aleothopteris* leaves from St. Clair, Pennsylvania. Unlike fern fossils from the Midwest and West, however, these plants often are preserved in black shale, where the contrast between the fossil and the rock is minimal. At times, these fern fossils can consist of complete fronds, some quite large.

Aleothopteris serli: from Middle Pennsylvanian strata of West Virginia. This genus often leaves distinct, clear impressions. It must have been somewhat stiff to have done so. (Value range G.)

Aleothopteris serli: West Virginia, light colored matrix (shale) is uncommon in the Appalachian Mountains, where plant fossils generally are grey or black. (Value range G.)

Coal seam adit intersected by a highway cut. The lower level of the coal seam is at the man's head. The coal bed is capped by a roof of thin sandstone. The following plants came from roof shale, which occurs just above coal seams in West Virginia. Appalachian Mountain coal beds have been honeycombed with mine tunnels like this. Coal mining techniques today can remove hundreds of feet of overburden, which can result in the removal of an entire mountain top, to get at coal seams six to eight feet thick like this one.

Pecopteris Fronds, Pennsylvania

Coal mining in Pennsylvania and eastern Ohio can uncover fossil ferns in shales overlying the coal. Large numbers of fronds of the genus *Pecopteris*, a genus which is representative of the Mid-Pennsylvanian, can be especially abundant. Often layered one on top of another, these are usually found in black or dark grey shale so that the contrast is not good; but, the specimens are significant in both their large number and in the completeness of the fronds.

Neuropteris frond impression: this is a typical fern fossil found in the roof shales of coal beds mined in western Pennsylvania. They are in black shale and the contrast between the fern and matrix is low. Often they are portions of the large fronds of common coal swamp genera. While often occurring in the roof shales of coal mines, they are seldom collected by miners from the mines, so that they are not as widely seen as might be expected considering their abundance. (Value range F.)

Pecopteris sp.: overlapping fronds, as in the previous specimen, which retain little original leaf material. Leaf material is usually preserved as a carbonaceous film. Lacking such original plant material, such slabs are less desirable than are those that retain the carbonaceous plant material as a compression. (Value range G.)

Pecopteris flexuosa: overlapping compressions of multiple fronds. These compressions retain some of the original leaf as a carbonaceous (coaly) film or residue. This makes them more attractive and desirable. Complete fronds are difficult to collect, even in artificial excavations like coal mines, as the shale tends to break into small pieces rather easily. It is difficult to retrieve the large slabs necessary to obtain a complete frond. (Value range F.)

Neuropteris flexuosa: partial frond or frond fragments. These can occur in large quantities in the roof shales of coal mines in Pennsylvania, as well as elsewhere. Those from Pennsylvania often look like this. They make nice specimens to introduce a young person to fossil plants as smaller ones can be both common and often are readily available. (Value range H.)

European Late Carboniferous Coal Swamp Floras

The European Late Carboniferous is essentially the same as the North American Pennsylvanian. Late Carboniferous coal swamp floras of Europe are also essentially the same as Pennsylvanian floras of North America. This might be expected as North America and Europe were once part of the same continental land mass, a land mass split and separating during the Mesozoic Era only after the existence of the coal swamps of Laurentia formed their coal. Laurentia refers to the large landmass that separated from the northern part of Pangaea in the Mesozoic Era as the Atlantic Ocean opened and then separated North America from Europe.

Pecopteris sp. (frond): Upper Carboniferous. Tremon (Leon), Spain. Preserved in slightly metamorphosed black shale in a manner similar to ferns found associated with anthracite coal in Pennsylvania. (All coal associated with such ferns would also be converted to anthracite.) *Courtesy of Steve Holley.* (Value range F.)

Neuropteris sp.: a typical coal-swamp fern in shale. Upper Carboniferous. La Magdalena (Leon), Spain. The European coal swamp flora is essentially the same as that of eastern North America. *Courtesy of Steve Holley.* (Value range G.)

Pecopteris sp.: flip side of the same slab as in above photo. Original plant material has been converted to a film of shiny graphite. Upper Carboniferous. Tremon Spain. (Value range F.)

Pecopteris sp. and *Cordaites* leaf. Upper Carboniferous. Puerto Ventura (Ashinas), Spain. (Value range F.)

Upper Pennsylvanian Plants of the U.S.

A number of Late Pennsylvanian fossil plants have come from eastern Kansas and the Kansas City area, as well as from Oklahoma. These fossils are sometimes different from the fossil plants of the coal swamps. There is, in fact, a gradual change in the coal swamp floras in the U.S. Midwest from lower to late Mid-Pennsylvanian. The differences between different species of the same genus which reflect this change are, however, often quite subtle. Because of the large multi-state area over which these floras extend, there has been a great deal of speciation on them, much more than is warranted in the author's opinion. Late Pennsylvanian floras, on the other hand, do appear to be more regionally diversified. Unlike earlier floras, late Pennsylvanian species of common genera, like *Pecopteris,* really can be distinctive. Late Pennsylvanian fossil plant localities are also not as widespread as are the earlier ones. This is partially because these younger strata seldom contain mineable coal seams in the U.S. By late Pennsylvanian time, the plants themselves are also suggestive of a less humid and warm environment. It appears that changes in the composition of the floras reflect this climatic change.

Permian floras represent a continuation of those of the late Pennsylvania, with dominant coal age plants like Calamites, *Lepidodendron*, *Sigillaria,* and many of the "fern" genera like *Neuropteris* becoming extinct with the Permian extinction event, arguably the world's most profound extinction.

Neuropteris cf. *N. flexuosa*: Upper Carboniferous, Westphalian, Brunssum, Netherlands. In the southern part of the Netherlands occur underground mines working the same coal beds found in Belgium. Paleozoic rocks do not occur on the surface in the Netherlands. All outcrops in that country are younger, either Cretaceous or Pleistocene.

Asterotheca sp.: a small frond preserved in what is known as burnout. Burnout can form during the spontaneous combustion of impure coal, often discarded at the mine. Plant bearing shale or clay lenses associated with such impure coal can then be fired or baked to form a brick-red rock that may contain plant impressions like this. Burnout is more commonly associated with geologically younger lignite or brown coal. It's not found too frequently associated with Paleozoic coals. Upper Carboniferous, Westphalian, Czevwionlier, Poland. *Courtesy of Steve Holley.* (Value range G.)

Neuropteris "Macroneuropteris" heterophylla or Neuropteris scheuchzeri: these are the Winterset "mystery ferns" of Chapter 9. Kansas City collector and amateur paleobotanist Timothy Northcutt believes these fossils are a species of *Neuropteris.* Genera and species of Late Pennsylvanian fern foliage are often distinctly different from that of the Middle and Lower Pennsylvanian and are also less commonly seen and collected. A large number of these ferns (labeled as *Pecopteris* sp.) came from near Independence, Missouri, and have entered the collector's market. Winterset Formation, Lower Upper Pennsylvanian, Jackson County, Kansas City, Missouri, area. (Value range G.)

Above: *Neuropteris(?)* sp. Left: unknown form, Winterset Formation. Lower Upper Pennsylvanian, Jackson County, Missouri, Kansas City, Missouri, area (Independence).

Same as the previous images: often labeled as *Pecopteris* sp., this fern is quite different from *Pecopteris*. It is probably a species of *Neuropteris*. Late Pennsylvanian floras can be quite different from those of the Lower and Middle parts of the Pennsylvanian Period.

Neuropteris(?) Late Pennsylvanian, Independence, Missouri.

Cyclopteris sp.: another Winterset mystery fossil. These broad leaves may have grown at the base of fronds of *Neuropteris*. However, this specimen is different from species of *Cyclopteris* of the Middle Pennsylvanian and may not even be that genus. Late Pennsylvanian plants can be different from the better known, Lower and Middle Pennsylvanian forms. *Courtesy of John Stade*

Neuropteris-like fern (top) and *Cyclopteris* sp. (bottom left): Winterset Formation, Independence, Missouri. (Value range F.)

Aleothopteris sp. Upper Pennsylvanian, West Virginia. (Value range F.)

Odontopteris brardii: Late Pennsylvanian ferns from the Bonner Springs Shale at Parkville, northwest Missouri. Numerous specimens of this attractive fern have been collected and disbursed by Kansas City collector Tim Northcutt (see Chapter 9.)

Aleothopteris serli in light grey shale: labeled as coming from Upper Pennsylvanian strata of West Virginia; it looks similar to *Aleothopteris* of the Middle Pennsylvanian. West Virginia has a wealth of fossil plants from the Pennsylvanian Period. (Value range G.)

Aleothopteris and lepidodendron leaves. Middle Pennsylvanian, West Virginia.

Pecopteris sp. frond. Upper Pennsylvanian, West Virginia. *Courtesy of St. Louis Science Center.*

Crossopteris sp.: a Permian fern. Rotlegendes, Lower Permian, Rhineland Pfaltz, Germany. This is a Permian fern from a sequence of Permian strata in Germany known as the Rotlegendes, a sequence of strata that, like Permian strata worldwide, contains large amounts of red beds. (Value range G.)

Silicified log from a tree fern: petrified wood from the Paleozoic is much less common than is wood found in strata of the Mesozoic and Cenozoic Eras. Petrified wood from these later eras is generally from either conifers or angiosperms, trees composed of wood more likely to be petrified (permineralized) as it was less "punky" and did not rot as quickly as that of most Paleozoic trees. Middle Pennsylvanian, Callaway County, Missouri.

Close-up of petrified log of a large tree fern. Middle Pennsylvanian, Callaway County, Missouri.

Tree-fern Petrified Wood and Coal Balls

Besides compressions and impressions of fern-like foliage, impressions and petrifactions of tree fern trunks can locally be common in Late Paleozoic strata. Petrifactions are when the wood of the plant is either replaced with minerals like quartz or is impregnated with minerals, as is the case of coal balls. This often is done at the molecular level so that details of the plant's cellular structure can be preserved with considerable detail.

Slice of a tree-fern log: the white areas are composed of quartz that has filled in parts of the log where the wood was not replaced with minerals because it quickly rotted away. The wood rotted out so quickly because large woody masses were "punky" rather than firm, like the wood of modern trees. Wood that rotted quickly left open areas that were later filled in with quartz, forming what suggests geodes. (Value range E.)

Slice of a tree fern log shown in the previous photo. Large, well preserved slabs of Paleozoic wood are relatively rare; however, petrified wood collectors often appear to be unaware of the significance of geologic time in their interest. Otherwise petrified wood like this would be more desirable among collectors than it is. (Value range E, polished round.)

Lens of river-deposited silty sandstone containing numerous fragments and impressions of tree-fern wood. This is essentially part of a fossil drift wood mass uncovered by I-70 improvement excavations, Cool Valley, Missouri, 2002.

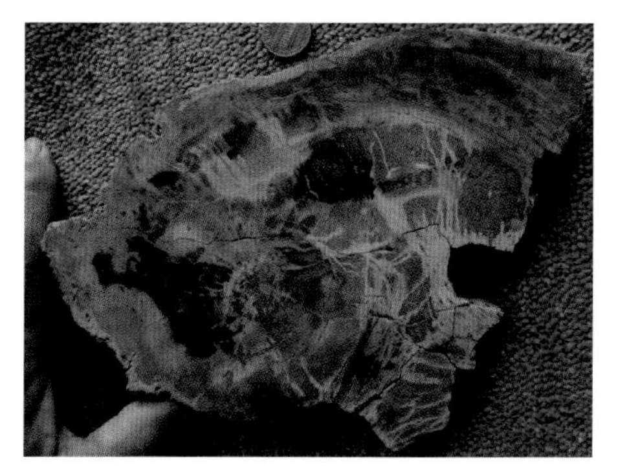

Slab of gnarly wood of a tree fern. Middle Pennsylvanian, St. Louis County, Missouri. (Value range F.)

Tree fern wood slice: notice that no trace of annular rings are seen in any of this wood. This slice came from a mass of petrified wood that, like much fossil wood of the Pennsylvanian, suggests large, gnarly tree ferns. The environment under which they grew was tropical. North America during the late Paleozoic was close to the equator, so that annular rings, which are produced from the changes of seasons, are absent. The climate was wet and hot all year round. Logs showing annular rings (both fossil and modern) are characteristic of higher latitudes. Middle Pennsylvanian, Callaway County, Missouri. (Value range F.)

Massive tree fern wood, Middle Pennsylvanian, Lincoln County, Missouri. Chunks of this brownish tree fern wood are found sporadically over large areas of central Missouri, especially in the hills just north of the Missouri River. It is also found in southern Illinois and Indiana, as well as in central Kentucky. Chunks of this petrified wood often are not recognized as being such, as it is often in the form of gnarly masses looking like many other rocks. Also Midwesterner's usually don't think of petrified wood being found in their area. They associated it with regions further west in the U.S., where it is much more common. Midwestern wood, however, is much older and rarer than the geologically younger petrified wood found elsewhere. This Paleozoic wood is preserved with quartz, which is capable of taking a high polish that is quite attractive. (Value range F.)

Tree fern outer bark impression in silty sandstone. Most late Paleozoic plants were punky (or reedy) and were not very strong. These punky logs leave impressions (in sand) like this. I-70 improvement excavations, Cool Valley, Missouri, 2002.

Tree fern, outer bark impression. Wabash Railroad cut, Cool Valley, Missouri.

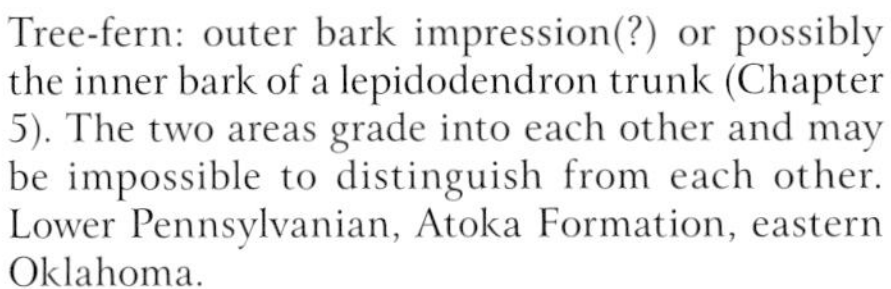

Tree-fern: outer bark impression(?) or possibly the inner bark of a lepidodendron trunk (Chapter 5). The two areas grade into each other and may be impossible to distinguish from each other. Lower Pennsylvanian, Atoka Formation, eastern Oklahoma.

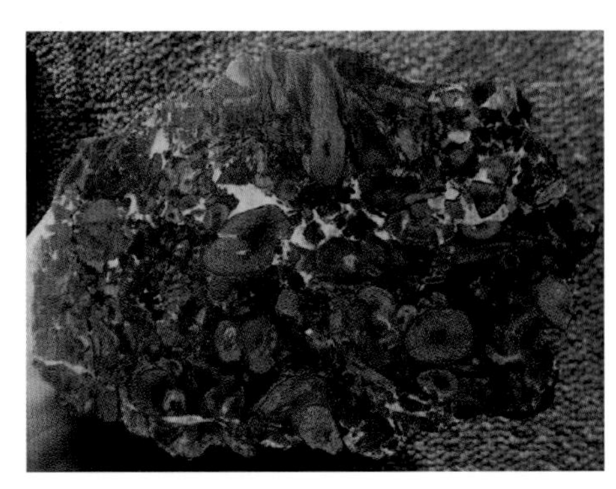

Cut slab made from above coal ball containing a mass of the intertwined root mantle of the tree fern *Psaronius*. Notice how well the roots and rootlets have been preserved in the coal ball. Coal balls form from the infiltration of calcium carbonate into what originally was a peaty mass. This mineral embeds the plant tissue so that it is retarded from being converted to coal. In this process, unlike in coal itself, plant tissue is preserved both chemically (as lignitic material) as well as being sealed and preserved in the calcium carbonate. This results in a plant fossil where tissue is preserved in such a way that it can be studied on a microscopic level in the same manner as done with modern plants. Berryville, Illinois. *Courtesy of Henry N. Andrews.*

A portion of a coal ball containing the buttressing roots of the tree fern genus *Psaronius*. Tree ferns were usually buttressed up in the coal swamps by large masses of roots that surrounded the central woody mass from which the foliage extended. These buttressing roots could be of two types, either intertwining roots or parallel ones. The intertwining type is seen here.

Another *Psaronius* slab showing buttressing tree-fern roots. The white material is calcium carbonate, which actually forms the coal ball. Coal balls can occur with mineable coal seams. They also weather from coal outcrops (especially in Illinois), where they can accumulate in creek beds that have cut through coal seams.

Psaronius cf. *P. blicklei*: this is a portion of the intertwined tree-fern root mantle preserved in a silicified coal ball. Here the petrifying material is silica rather than the more common calcium carbonate. This petrified mass was found as a glacial erratic. It was plucked from an outcrop, probably from somewhere in Illinois, and transported south by glaciers to eastern Missouri during the Pleistocene ice age. (Value range F.)

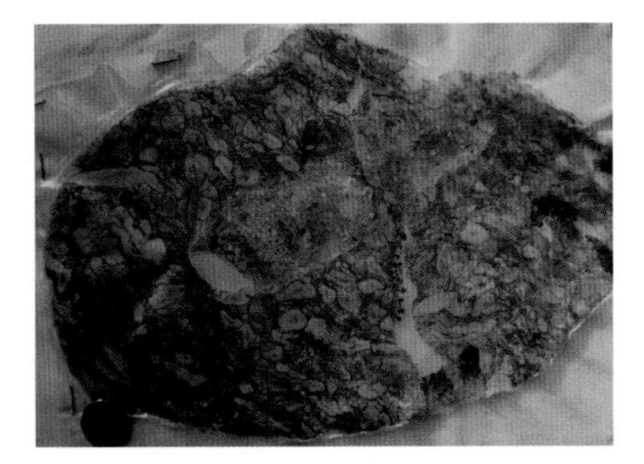

Mass of *Psaronius* (tree fern) root mantle preserved in a coal ball. The grey center (inside the coal ball) is limestone. This is an acetate peel made by an interesting transfer process from an acid etched coal ball. Middle Pennsylvanian, near Berryville, Illinois. *Courtesy of Henry N. Andrews*

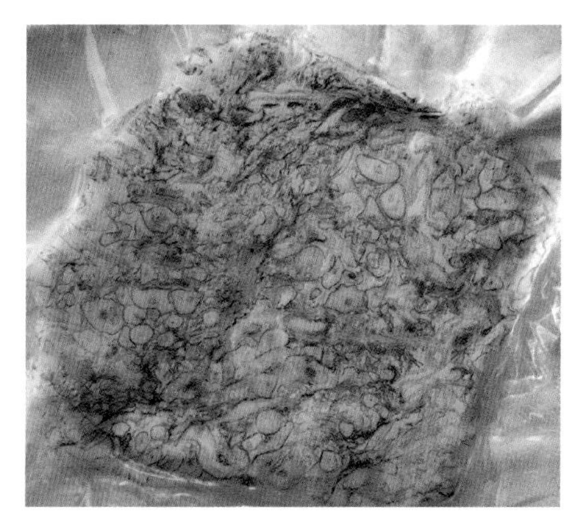

Coal ball peel showing intertwining root mantle of a coal swamp tree-fern. Berryville, Illinois.

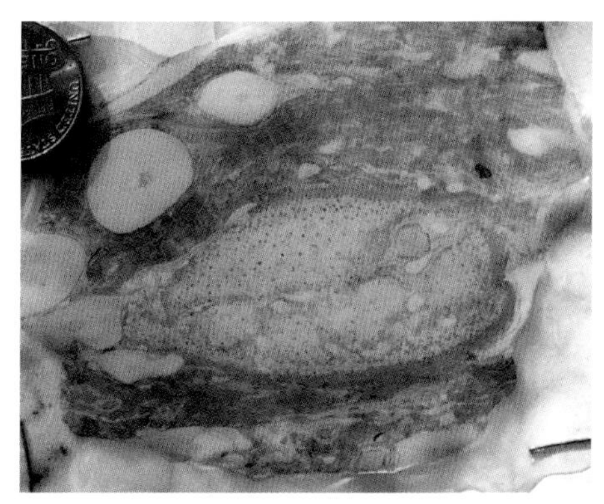

Petiole of a large tree fern sectioned in a coal ball acetate peel. Berryville, Illinois.

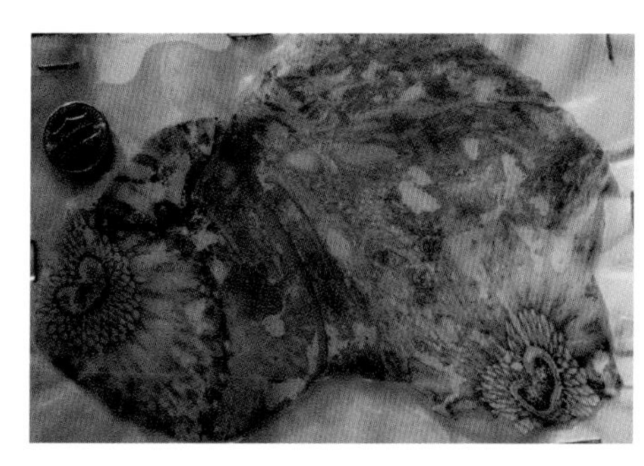

Psaronius blicklei (two specimens) showing buttressing root mantle in which the roots are arranged in a parallel and regular manner.

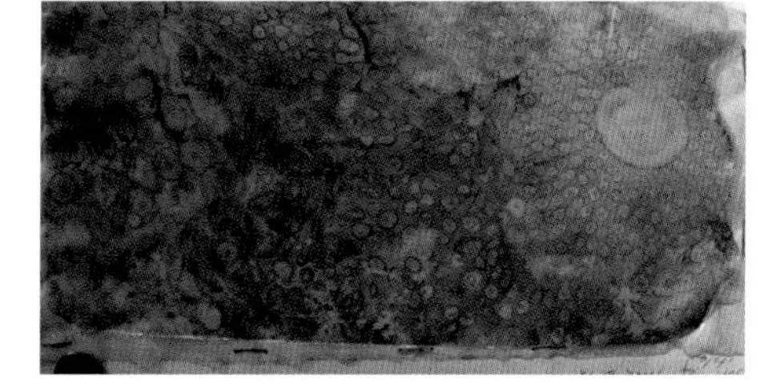

Massive buttressing root mantle (arranged in a parallel manner) of *Psaronius blicklei* with a portion of the trunk showing a stele. Middle Pennsylvanian, Berryville, Illinois. *Courtesy of Henry N. Andrews*

Psaronius sp.: a nice section of this tree fern exposed in the interior of a sliced coal ball. Middle Pennsylvanian, Berryville, Illinois. *Courtesy of Henry N. Andrews*

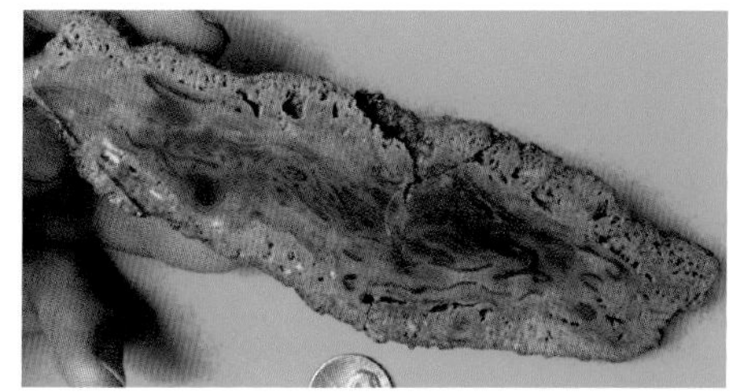

Psaronius sp.: silicified specimen of a *Psaronius* tree fern from Callaway County, Missouri.

Petrified Tree Fern Wood from Brazil

Numerous and colorful slices of petrified tree fern logs have come onto the fossil market from the Permian of Brazil and these have become widely distributed among collectors, most coming through the Tucson show. Most of these come from the Lower Permian Pedra de Fogo Formation in the Brazilian state of Tocantins in the region between Araguaina and Filadelfia, Brazil. They are both colorful and attractive.

Psaronius brasiliensis (*Tieta singularis*): an especially symmetrical slice of a common Permian tree-fern from Gondwanaland sediments of Brazil. A large number of polished slices of these tree-fern trunks have entered the fossil collector's market, most coming through the Tucson, Arizona, show. They usually show the buttressing root mantle surrounding the stele of the tree fern itself. Araguaina or Filadelfia, northeastern Brazil. (Value range F.)

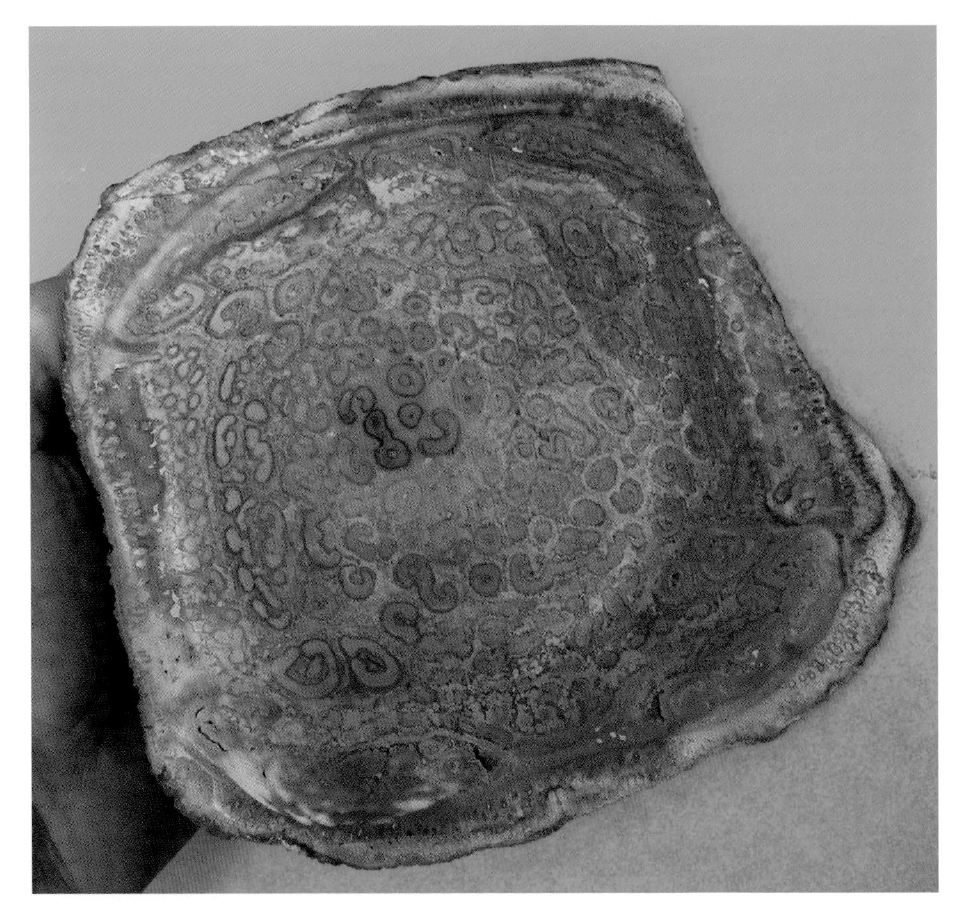

Less symmetrical round of this attractive Paleozoic plant. (Value range F.)

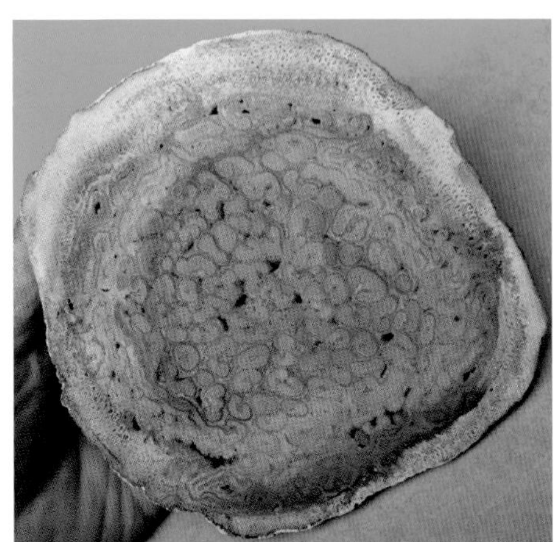

(Value range F.)

An especially symmetrical round. Araguaina, northeastern Brazil

(Value range F.)

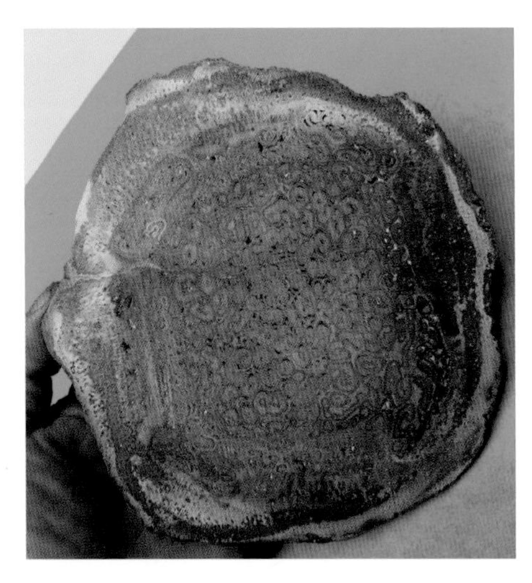

(Value range F.)

Bibliography

Wittry, Jack, 2006. *The Mazon Creek Fossil Flora.* Downers Grove, Illinois: Esconi Associates.

Glossary

Burn out: shale or clay above a coal seam fired or baked to a brick-red color from the spontaneous combustion of associated coal. Burnout sometimes can contain leaves or other foliage that, although having no original plant material, can often make attractive plant fossils.

Laurentia: this consists of most of the North American continent, which formed by separation from Europe and Asia after the Paleozoic Era. Because both eastern North America and Europe in the Carboniferous, at times, were covered by coal swamps, the flora and fauna of this time is almost identical in these two regions.

Linnean names: scientific names of organisms (including fossil organisms) in which the name is a binomial (genus and species). This system of nomenclature was first proposed by eighteenth century Swedish naturalist Carl Linne (Carlos Linneaus) and is known as the Linnaean System of Nomenclature. Names like *Pecopteris grandifoli* would be an example, *Pecopteris* being the genus and *grandifoli* the species. With fossil plants, different Linnaean names are often given to different portions of the same plant, such as foliage, seeds, and trunks, as these are usually found separately in the fossil record.

Permian Extinction Event: the end of the Permian Period marks the end of the Paleozoic era; it also marks the end of the greatest number of life forms recorded in the rock record. This is considered to be the most profound extinction event in earth's geologic history.

Tucson Show: a large fair spreading over various parts of Tucson, Arizona, featuring geo collectibles such as minerals, rocks, meteorites, and fossils. This show takes place in late January and early February. A wide variety of fossils (including fossil plants) are dispersed by dealers and collectors through this show.

(Value range F.)

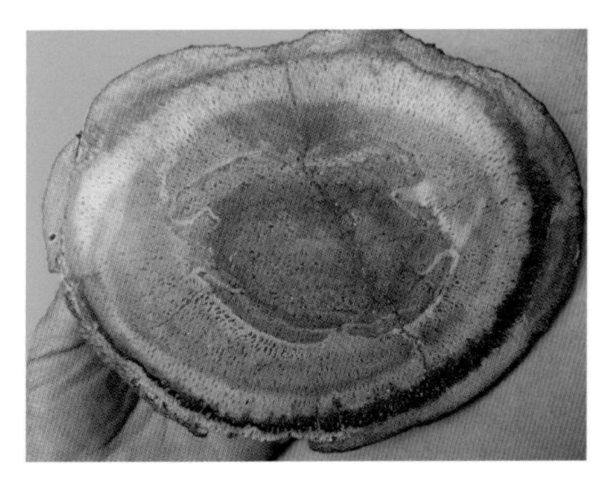

(Value range F.)

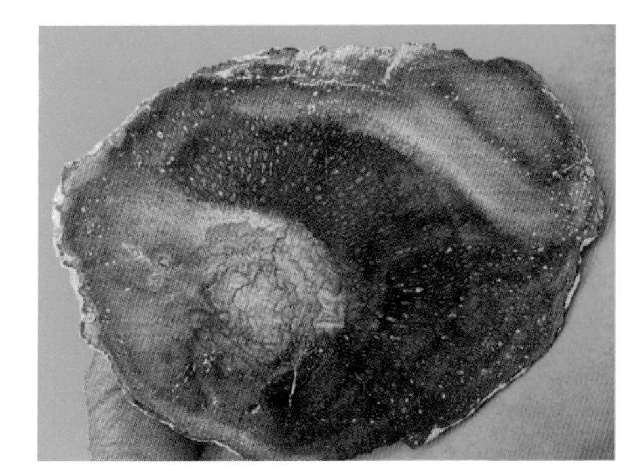

A less symmetrical specimen suggestive of *Psaronius* of the northern hemisphere.

Chapter Five
Scale Trees (Lycopodophytes)
Giant Prehistoric "Snakes"

These prehistoric trees have often confused more people than any other fossil plant. These fossils resemble impressions of what look like large, scary, prehistoric snakes! These peculiar plants, often referred to (because of this snake-like appearance) as "scale trees," may have had their first appearance in the Silurian Period(?) (or lower Devonian) as an odd fossil plant found in New Zealand known as *Baragwanathia*. Whether it truly is a member of this plant division, the Lycopodophytes are definitely one of the older groups of land plants. The first undoubted "scale trees" appear in the Devonian and in the Mississippian (or the Lower Carboniferous) they became one of the most widely found fossil land plants of this part of geologic time.

Lycopodophytes are plants where the entire plant is covered with elongate leaves that, when removed from the trunk and branches, leave behind scale-like leaf scars. Lycopodophytes of the Paleozoic were often large plants, plants which had their maximum diversity during the Mississippian and Pennsylvanian Periods (Carboniferous of Europe). Today they are represented only by herbaceous plants that grow in cool, wet regions and are known as lycopodium, from which the Divisions name is derived. Extinct lycopods consisted of two major groups: the **Lepidodendrons** and the **Sigillarids.**

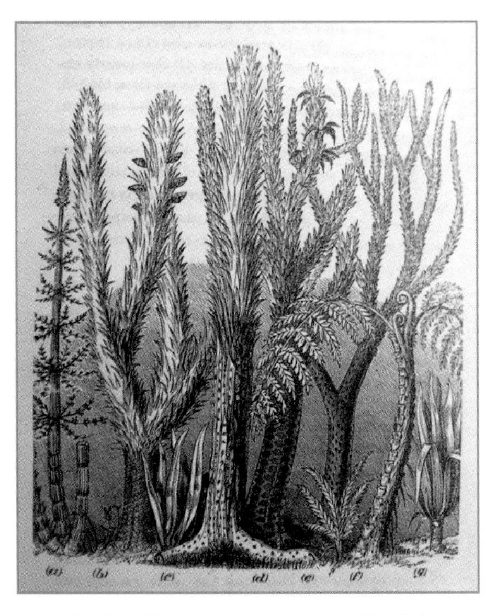

Late Carboniferous reconstruction-II. *Calamites* are at the far left, various types of *Sigillaria* are in the middle, and *Cordaites* are at the right. From J. William Dawson's, *The Story of the Earth and Man.*

Pennsylvanian or Late Carboniferous coal swamp reconstruction-I. Lepidodendron "trees" are at the left, Arthrophytes (Calamites) at the right. *Painting courtesy of Dorothy J. Echols*

Mississippian (Lower Carboniferous) Lepidodendrons

Lepidodendrons were generally relatively large plants. They were covered with (usually) elongate leaves, the arrangement of which occurred diagonally wrapped around the plant's trunk and branches. Both lepidodendron and sigillaria have distinctive, dichotomizing roots, which leave root casts known as *Stigmaria*. Lepidodendron was especially well represented and characteristic of the Lower Carboniferous or Mississippian Period (as it is known in the U.S.). Unlike in the Upper Carboniferous (Pennsylvanian of the U.S.), coal seams are rare in the Mississippian, although its sediments are sometimes laden with black carbonaceous (coal-like) matter, especially the shales. By far the most common plants in Lower Carboniferous strata are impressions of the trunks of various types of lepidodendron and related lycopods. Lepidodendron is often associated with sandstone beds of the Lower Carboniferous, particularly sandstones of the later portion of the period, which in the U.S. is referred to as the Chesterian Series, a name derived from outcrops along the Mississippi River near Chester, Illinois.

Lepidodendron sp.: an impression of distinct leaf cushion impressions in a sandstone concretion. This specimen came from a (presumed) water logged lepidodendron stem washed into the rapidly subsiding trench of the Ouachita Geosyncline. Jackfork Sandstone, Upper Mississippian, Ouachita Mountains, Arkansas. (Value range E.)

Knorria sp.: the inner bark (inner cortex) of a lepidodendron branch. Different parts of a lepidodendron trunk are given specific generic names. This is a consequence of these fossils originally being considered as different types of trees rather than different layers of the same tree. Cypress Sandstone, Chester Series, Upper Mississippian, Chester, Illinois. (Value range F.)

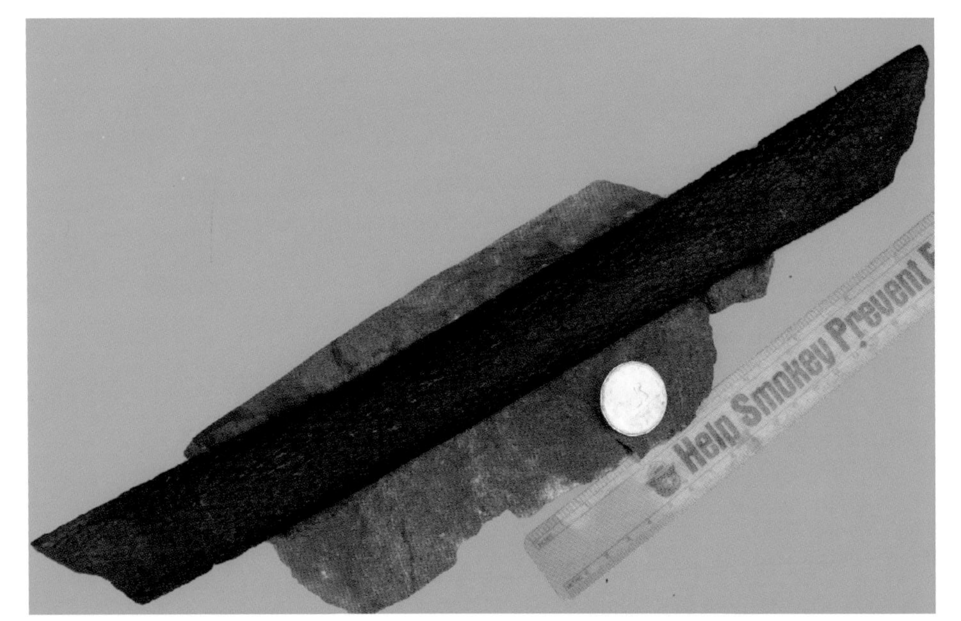

Lepidophilos: Mississippian (Lower Carboniferous) scale tree from near St. John, New Brunswick, Canada. Various genera and species of lepidodendron-like plants are most common and dominant in the Lower Carboniferous. In the Lower Carboniferous, they are generally medium-sized forms; larger forms are more common in the Upper Carboniferous. This is a pith cast of this genus of medium-sized scale tree.

Lepidodendron sp.: in the U.S. Midwest, lepidodendron is the most commonly found land plant in Mississippian age strata. This is an impression of its scale-like surface of leaf scars in sandstone. Tar Springs Sandstone, Upper Mississippian, Star Landing, Missouri. (Value range F.)

Lepidodendron sp.: an impression of a compressed medium-sized lycopod branch in silty limestone. Land plants are uncommon in limestone. Salem Formation, Middle Mississippian, St. Louis County,

Stigmaria, Root Casts of Lepidodendron and Sigillaria

Stigmaria is often associated with what is known as underclay, sediment that formed the soil directly underneath a coal swamp. Usually specimens in this paleosoil (or paleosol) are poorly preserved and vague. The best specimens of Stigmaria are found as casts in sandstone.

Stigmaria sp.: a section of one of the roots of a large lycopod, probably Lepidodendron. These can be fairly common fossils, usually found in sandstone. This is a sandstone filling of what originally was "reedy and punky" wood. (Value range F.)

Stigmaria sp.: two sections of roots of either *lepidodendron* or *sigillaria*, Lower Pennsylvanian, Arkansas. Stigmaria is often associated with coal underclays—the soil upon which a coal forest grew. When found in this manner they are usually poorly preserved.

Representative Lepidodendron Fossils

Lycopodophyte fossils occur in a variety of forms, especially in Pennsylvanian (Upper Carboniferous) strata. Those preserved in fine-grained sandstone, like these shown here from Iowa, can be especially clear and attractive.

Lepidodendron giganteum Lesquereux: a species of Lepidodendron with exceptionally large leaf scars. This specimen is silicified with an internal structure suggestive of a tree fern (*Psaronius*). It came from the bed of a southwestern Iowa stream, which drains into the Grand River of northwestern Missouri. Numerous silicified specimens of this species, with exceptionally large leaf scars, have been found in the bed of this stream. Upper Pennsylvanian, Clorinda, southwest Iowa. (Value range F.)

Lepidodendron giganteum Lesquereux: the large leaf scars of this species are especially distinct on this specimen from near Clorinda, Iowa.

Lepidodendron vestitum: impressions of Lepidodendron fragments in hard, slaty shale from above an anthracite seam, eastern Pennsylvania. Specimen from St. Louis Academy of Science.

Lepidodendron sp.: specimen in slaty shale from an anthracite mine in eastern Pennsylvania. (Value range G.)

Lepidophyllum sp.: elongate leaves of the Lepidodendron. These leaves were attached to the trunk and branches of the Lepidodendron—most were elongate like these. Upper Pennsylvanian, West Virginia. (Value range G.)

Lycopodophyte Fossils from the Famous Braidwood (Mazon Creek) Fossil Beds of Northern Illinois

Ironstone concretions from the Mazon Creek fossil beds of northeastern Illinois contain a variety of fossil lycopodophytes. Here are some representative examples.

Spoil piles, Coal City, Illinois, 1953. These mounds of shale and soil were produced from surface coal mining in the 1940s. In the 1950s, they had not yet vegetated and fossil bearing nodules (or concretions) that weathered out of them had not yet been combed over by many collectors.

Portion of a large trunk of a Lepidodendron in grey shale of the Francois Creek Formation, Braidwood, Illinois. *From the John McLuckie collection*

Three dimensional, small (terminal(?)) lepidodendron completely removed from an ironstone concretion. (Value range G.)

Small branch of *Lepidodendron*, (*L. obovatum*). (Value range G.)

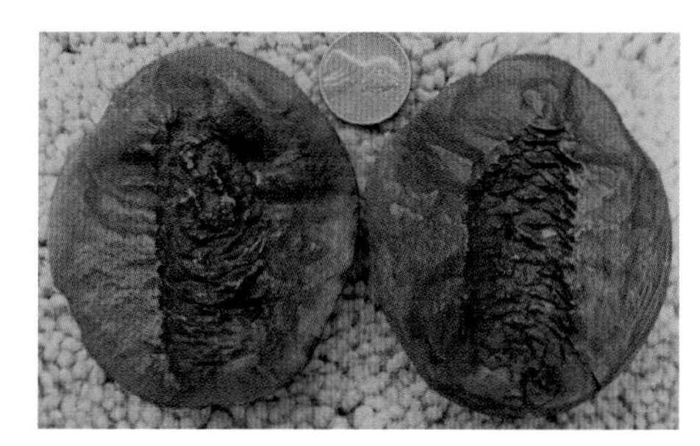

Lepidostrobos sp.: impression of a lepidodendron "cone" or reproductive body. (Value range F.)

Lepidophloios sp. Lesquereux: a lepidodendron with large, protruding leaf scars. (Value range F.)

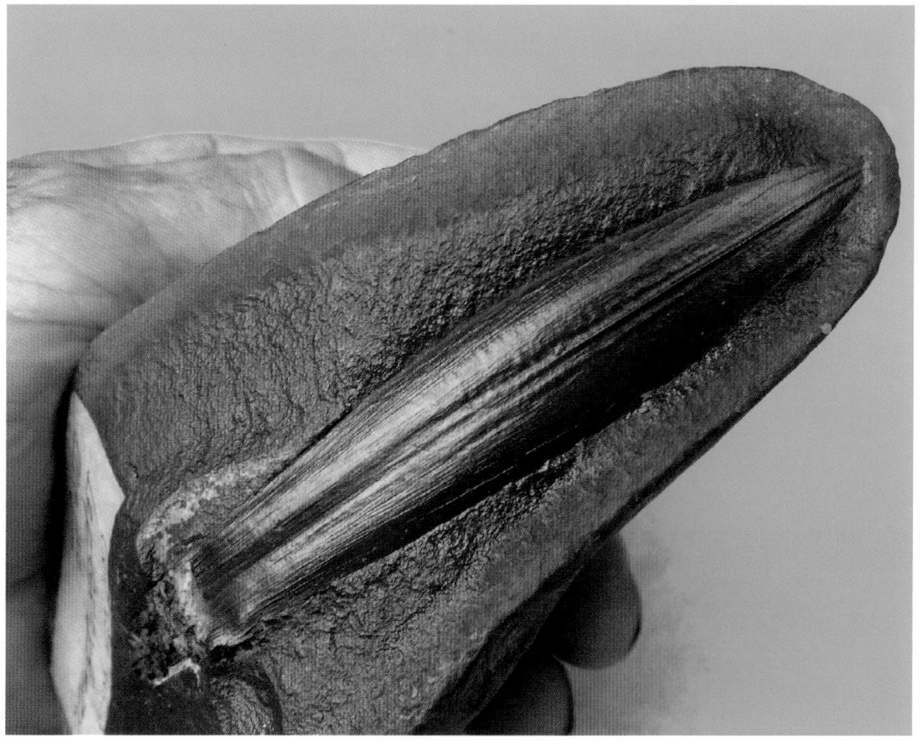

Lepidophyllum majus Brongniart: a short, somewhat blunt exquisitely preserved lepidodendron leaf. (Value range E.)

Lepidodendron intermedium: typical leaf of the lepidodendron (part and counterpart). This is the most frequently found small lepidodendron leaf. (Value range G.)

Lycopodites sp.

A small, herbaceous fossil lycopod.

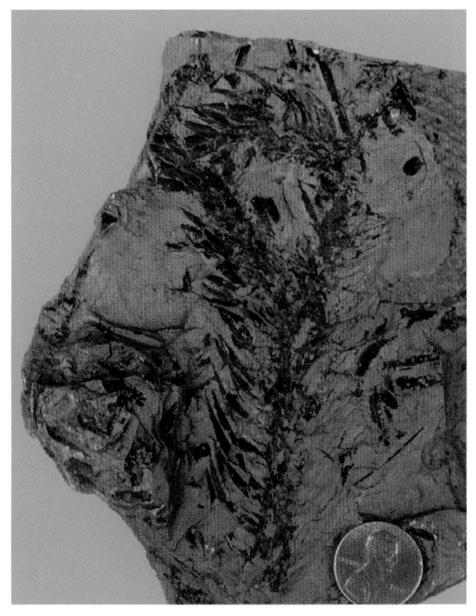

Lycopodites sp.: compression of a small lepidodendron from the same locality as the previous fossil. Windsor, Henry County, Missouri. (Value range F.)

Lepidodendron aculeatum (left) with *Pecopteris* sp., partial fronds. From the same locality as the previous *Lycopodites* specimens. Windsor, Henry County, Missouri. (Value range F.)

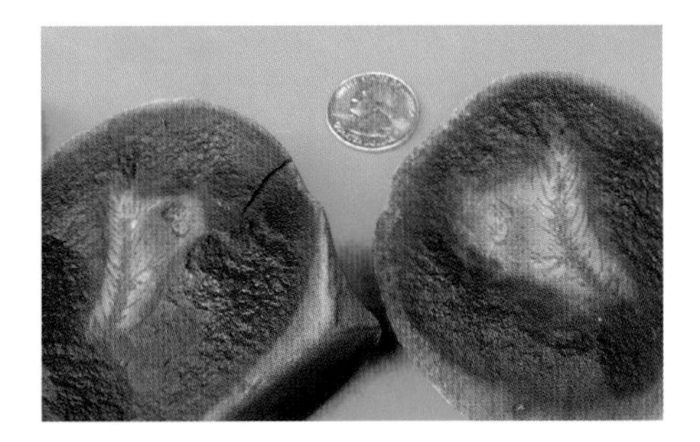

Lycopodites meeki: a small, herbaceous lycopod essentially like the living lycopodium.

Lepidophyllum longifolium: a scattering of these elongate leaves is a fairly common phenomena in Pennsylvanian plant bearing horizons, especially in the roof shales overlying coal beds. They probably are the shed leaves of the lepidodendron and are similar to leaf litter found in today's forests and swamps. They are larger and longer than the leaves of Lycopodites. Upper Pennsylvanian, West Virginia.

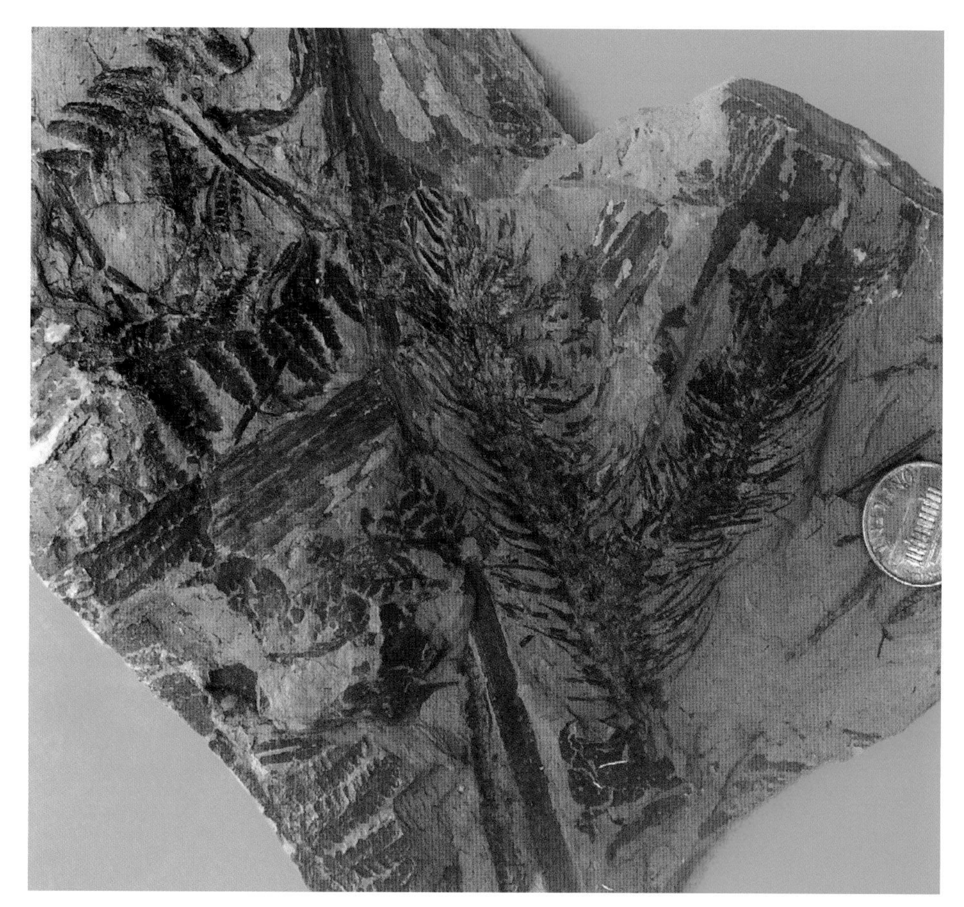

Lycopodites sp.: compression of a diminutive lycopod or what could also be the terminal portion of a larger lepidodendron tree. Croweberg Formation (coal) near Windsor, Henry County, Missouri. (Value range F.)

Lepidodendron Bearing Lower Pennsylvanian Sandstones

Thick Lower Pennsylvanian sandstones (Atoka Formation) of the southern Ozarks (Boston Mountains of Arkansas) and a thicker sequence of strata to the south deposited at what at the time was the edge of North America in the late Paleozoic. This strata may contain excellent lepidodendron specimens, many carried by rivers as water-logged driftwood flowing into what was, at the time, a subsiding trench. Strata that correlate with the Atoka Formation of Missouri, Iowa, Illinois, and Indiana, also yield similar lepidodendron specimens.

Sandstone impression of Lepidodendron trunk section in a road cut in the Millstone Grit of the Atoka Formation from the area shown in the previous photograph. Jasper, northwest Arkansas.

Thick plant-bearing sandstone beds of Lower Pennsylvanian age make up the Boston Mountains of the southern Ozarks in Arkansas. The millstone grit can be seen as the horizontal band in the upper part of the hill to the left.

Cross-bedded sandstone of the Atoka Formation (Millstone grit). Boston Mountains, northern Arkansas.

Lepidodendron sp.: small specimen found along the Ozark Highland Trail, Ben Hur, Boston Mountains, Arkansas.

Lepidodendron obtusum: Lower Pennsylvanian Sandstone, northeastern Missouri.

Lepidodendron obtusum Lesquereux: from a correlative of the Atoka Formation, Lower Pennsylvanian, Tennessee, Illinois. (Value range F.)

Knorria sp.: the inner bark of a Lepidodendron tree. Lower Pennsylvanian, southern Indiana. *Courtesy of John Stade* (Value range F.)

Asolanus sp.: a distinctive type of lepidodendron given its own generic name. It has distinct extensions of the leaf scars (seen in this cast of a large trunk as holes). Atoka Formation, Boston Mountains, Arkansas. (Value range F.)

Lepidodendron obtusum Lesquereux: impression in fine grained sandstone showing branching of the plant near its top. Lower Pennsylvanian, Pella, Iowa. (Value range E.)

Canadian Maritimes Upper Carboniferous

Lycopodophyte fossils of the Canadian Maritimes are almost identical to those of Europe, especially those of England and Wales. This is because what today are the Canadian Maritimes, before the opening up and formation of the Atlantic Ocean, were connected to what is now Western Europe.

A thick sequence of tilted Lower Carboniferous strata that form sea cliffs in the Bay of Fundy near Joggins, Nova Scotia, in the Canadian Maritimes.

Lycopodites sp.: a small, herbaceous lycopodophyte similar to the modern lycopodium. Joggins, Nova Scotia.

Knorria sp.: small stem showing the inner cortex of the lepidodendron plant. Joggins, Nova Scotia, Canada.

Sigillaria tessellate Brongniart: the other common, large scale tree in which the leaf scars are arranged vertically on the plant's trunk. South Joggins, Nova Scotia.

Lycopodophytes Preserved in Coal Balls

Coal balls are concretions composed of calcium carbonate, a mineral which formed when the plant material of a coal swamp was embedded in calcium carbonate **before the process of** coalification took place. Coal balls are randomly sliced and when a significant fossil is found, the cut surface is smoothed and then etched with dilute acid. A thin portion of the plant material thus exposed can then be embedded in a plastic sheet by an ingenious process discussed in the following chapter. Thin slices of a coal ball can also be mounted on a glass slide, ground down to a thin layer, and then studied under the microscope in the same way one would study modern plants.

Lepidostrobus sp.: portion of a lepidodendron cone preserved in a coal ball. A portion of an acetate peel of the specimen can be seen to the right. Mid-Pennsylvanian, Berryville, Illinois. (Value range F.)

Lepidodendron in a coal ball preserving leaf cushions as they might occur in a living plant. Middle Pennsylvanian, Berryville, Illinois, coal ball locality.

Cross section of a lepidodendron stem showing pith and radial attachment areas of the leaves of the plant. Coal ball peel, Berryville, Illinois. *Courtesy of Henry N. Andrews*

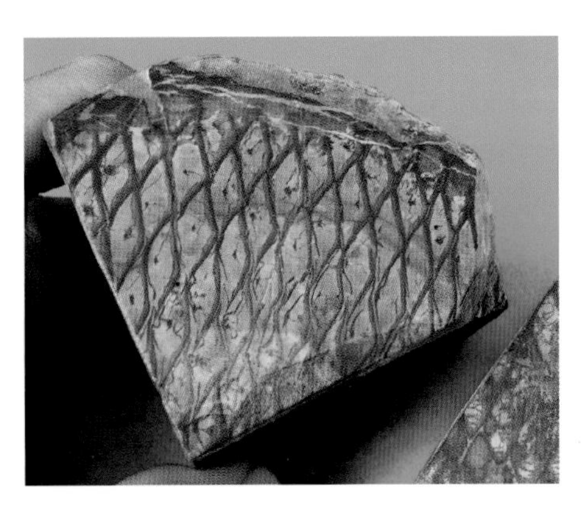

Close-up of a portion of a coal-ball preserved Lepidodendron.

Lepidodendron stem preserved in a coal ball and mounted on a glass slide. Such a paleobotanical preparation can be studied in the same manner as can a slide made from modern plant tissue. Middle Pennsylvanian, Berryville, Illinois.

Sigillaria

Sigillaria is the other large lycopodophyte associated with coal swamps. It has leaf scars arranged in rows that were parallel to the plant's trunk and branches, instead of being spirally arranged around them as is the case with lepidodendron. Leaves of Sigillaria are indistinguishable from those of lepidodendron.

Sigillaria tessellata Brongniart: a common form of Sigillaria. Braidwood (Mazon Creek) fossil beds, Braidwood, Illinois. (Value range G.)

Sigillaria sp.: a portion of the Sigillaria trunk with plant material present as coal, preserved in an ironstone concretion. Braidwood fossil beds. (Value range F.)

Sigillaria scutellata Brongniart: the distinct grooves (preserved here as ridges) are characteristic for this species. Braidwood fossil beds, Braidwood, Illinois.

Sigillaria laerigata Brongniart: a small portion of a large Sigillaria tree preserved as a sandstone impression. Carbondale Group, Middle Pennsylvanian, Sparta, Illinois.

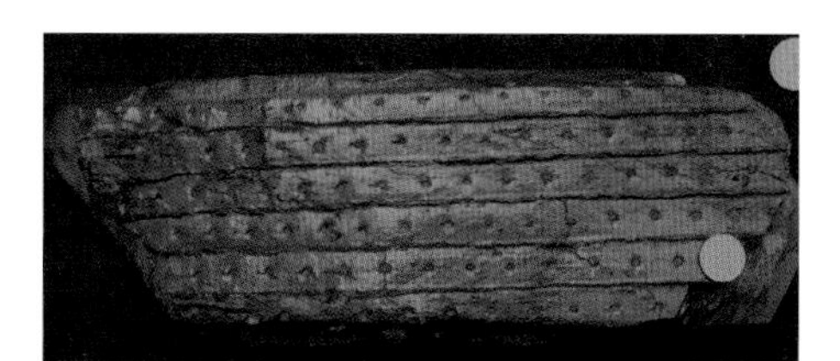

Sigillaria orbicularis Brongniart: impression (in shale) of a common species of Sigillaria. Middle Pennsylvanian, Dade County, Missouri. (Value range E.)

Sigillaria sp.: distinct specimen from outcrops exposed along the Grand River of the Warrensburg-Moberly Sandstone near Bedford, Missouri.

High wall of a coal strip mine in western Dade County, Missouri, 1968. The previously shown Sigillaria specimen came from this locality.

Syringodendron sp.: subsurface of a portion of the trunk of a Sigillaria tree. Inner portions of Paleozoic plants sometimes are given generic and specific names as they differ from the outer surface of the plant. Warrensburg-Moberly Sandstone, north St. Louis County, Missouri.

Outcrop of the Warrensburg-Moberly Sandstone (a Pennsylvanian river channel deposit) exposing a fossil driftwood mass from which the previous specimen was collected.

Valmeyerodendron

Limestone usually does not contain fossil land plants. Even geologically young limestone formed when land plants became abundant in the Mesozoic and Cenozoic Eras, rarely contain them. This is because limestone generally is deposited far from land, too far for most plants to be transported to where it formed. With this in mind, it is interesting to note that zones of land plants, often consisting of lycopods, occur with some regularity in the Middle Mississippian Salem Limestone of the U.S. Midwest. The Salem Formation is (in part) a pure, white fossiliferous limestone, the plants occurring with normal marine fossils like brachiopods and bryozoans. The most widespread land plant occurring in various plant horizons of the formation appears to be species of the genus *Valmeyerodendron*. This is an herbaceous lycopod unique to the Mississippi Valley region for which the Mississippian Period was named. *Valmeyerodendron* was described from the Salem Formation by James J. Jennings, who first described the genus. In addition to *Valmeyerodendron*, plants suggestive of cordaites also occur in the Formation and are covered in Chapter Seven.

Unidentified land plants (possibly a type of cordaites) or a lycopodophyte in limestone of the Middle Mississippian Salem Formation. St. Louis County, Missouri.

Close-up of land plants in limestone, a not-too-common occurrence in Paleozoic or other age limestone.

Valmeyerodendron sp.: part and counterpart of a Mississippian lycopodophyte. Salem Formation, Middle Mississippian, St. Louis County, Missouri. (Value range F.)

Valmeyerodendron sp.: impression (with no plant material preserved) of the trunk with leaf scars in limestone. Salem Formation, St. Louis County, Missouri. (Value range G.)

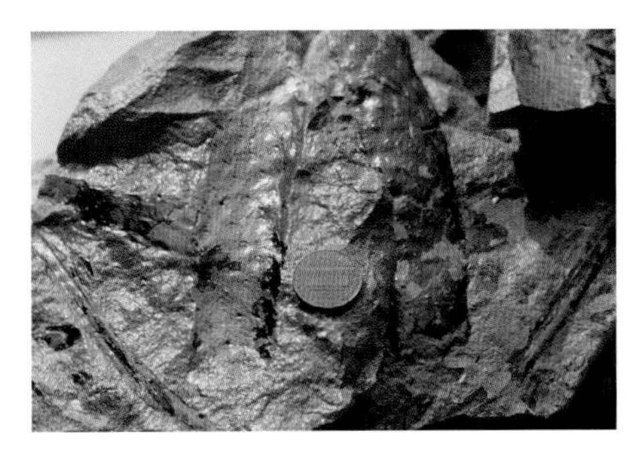

Valmeyerodendron sp.: a Mississippian genus of an undoubted lycopodophyte found in silty limestone layers of the Salem Formation, St. Louis County, Missouri.

Valmeyerodendron sp.: specimen with plant tissue preserved as a thin layer of coal.

Reconstruction of the Mississippian lycopodophyte *Valmeyerodendron*: it was a small herbaceous plant. It's peculiar in that it is always associated with Mississippian limestone. *Courtesy of James Jennings*

Carnotite (a uranium mineral) associated with plant bearing, carbonaceous silty limestone of the Salem Formation. Concentrations of plant matter in rock strata can attract and harbor uranium compounds to form concentrations of that radioactive element. The plant bearing zones of the Salem Formation as a consequence of this are often slightly radioactive.

Species and Genera of Lepidodendron and Sigillaria

Earlier literature on these plants generally used Lepidodendron and Sigillaria as a genus—variations of which were then given species names. As with so much of paleontology, the inevitable splitting has taken place, individual species becoming genera and greater attention being paid to slight morphologic variations, which then become species. This trend toward splitting can be confusing to the novice, who may see the same fossil given different names in publications that span only a few decades. The strategy used here on both plants is either to utilize the name given in the paleobotanical literature on that particular plant during the past two decades or to utilize what is given on the fossil label if the specimen was acquired from a reliable source. In recognition of this recent splitting, the more recent generic names are placed in parenthesis between the genus and species when such current taxonomic designations are available. Some might criticize this strategy; however, in a broad work such as this (as well as with other Schiffer works by the author) it is virtually impossible to know all taxonomic changes on a particular fossil that have taken place. **This is the reason why paleontology (and paleobotany) becomes so specialized**—specialists also should try to avoid excessively sharp criticism and realize that a general work such as this creates interest in their discipline, which generally is beneficial to all.

Bibliography

Jennings, James, 1971. A New Lycopod Genus from the Salem Limestone, (Mississippian) of Illinois. MS thesis, University of Illinois, Chicago.

Glossary

Carboniferous and Pennsylvanian Periods: in much of the United States and in western Canada, strata that belong to the Carboniferous Period of Europe are found to be made up of two rock sequences quite different from each other in both their sediment type and their contained fossils. The rock sequence of the early or lower Carboniferous (Mississippian Period) usually being limestone containing marine fossils; the sequence of the late (or Upper Carboniferous) contains more shale and sandstone, sometimes with fossil land plants. A hiatus or unconformity often separates these two different sequences. In the Canadian Maritimes, however, these two sequences are not so recognizable. This is the same situation in Europe, where what are collectively known as Carboniferous strata are not separated by any distinctive change.

Correlation of strata: the matching of rock strata deposited in different regions on the basis of that part of geologic time when such strata was deposited. This generally is done by the use of matching contained fossils or by matching similarities of their contained fossil flora and fauna.

Use of the Internet and fossil plants: by the use of both older and more recent names of fossils, a person who wants to acquire more information on a particular fossil can "Google" that name, generally with good results. An Internet search is an effective method for acquiring additional information, as fewer sites dealing with fossil plants are present than with animals. These Internet sites also appear to be more accurate than some sites dealing with animals. This is especially the case with dinosaurs.

Chapter Six
Arthrophytes (Calamites)

Giant scouring rushes—large reed-like plants that dominated Late Paleozoic landscapes—were common land plants, especially during the Pennsylvanian Period. Today, small, herbaceous plants known as scouring rushes are their descendents. All of these are known as arthrophytes (segmented plants) as their stems are partitioned into segments.

The most common Pennsylvanian arthrophyte is known as calamites. It's a reedy, often large plant, with a hollow pith that commonly became filled with sediment to yield what are distinctive segmented pith casts.

Like most large, late Paleozoic plants, the trunks of calamites (*Calamites*), foliage (*Annularia*), and reproductive structures (sporangia known as *Calamostachys*) are given separate names as is it rarely clear what portions of a specific plant went with what. Calamites reached their zenith during the Upper Carboniferous, where they occurred in extensive swamps in such large quantities as to form beds of coal.

Equisetum hyemale (scouring rush): these living plants are like diminutive calamites. Members of a once dominant and widespread division of plants, the Division Arthrophyta, *Equisetum* today is limited to localized, wet areas like swamps, road ditches, or lowlands close to permanent streams. These are growing along a trail that was formerly a railroad right-of-way. The area remains permanently wet because it is underlain by shale, which blocks descending groundwater moving through the Pennsylvanian (and calamites bearing) Warrensburg-Moberly Sandstone in northern St. Louis County, Missouri. The soil covered sandstone just above the impervious shale layer thus stays permanently wet, allowing the modern *Equisetum* to flourish.

Coal swamp reconstruction. Art is the only way by which a prehistoric scene can be "resurrected." This lithograph shows calamites, the large arthrophytes of the late Paleozoic (especially those of the Pennsylvanian or Upper Carboniferous) in the foreground. To the left, in the background, are large scale trees. From an early twentieth century Russian work courtesy of Dorothy J. Echols.

Group of modern *Equisetum hyemale*.

Reproductive structure (strobilus) of modern *Equisetum, E. aruense.*

Calamite-bearing Channel Sandstones (St. Louis Area *Calamites*)

In the northern portion of the St. Louis metropolitan area occur ancient stream sediments containing what are occurrences of fossilized driftwood preserved in beds of tan sandstone. These sandstones were deposited in ancient shifting river channels and appropriately are referred to as channel sandstones. As is characteristic of river deposits today, they contained a common component of river sand, driftwood. This fossil driftwood consists of impressions and casts of reedy plants, most probably derived from coal swamps that occurred adjacent to the river channels. As can be typical in an urban area, outcrops of these ancient channel deposits have come and gone. The channel deposits were (or are) exposed in creek beds and banks, but in many cases are only temporarily exposed by digging associated with construction.

Equisetum sp. Diminutive form of modern plant.

Fossil calamites collected from a concentration of them occurring in a fossil driftwood "pocket." From the I-70 rerouting excavations, north St. Louis County, 2002. This occurred close to the area where the previously shown modern *Equsitum* grows.

Another view of this (now covered) fossil driftwood pocket.

Fossil driftwood "pocket" (where people are) in Warrensburg-Moberly Sandstone, an ancient river deposit that lies above grey shale layers.

Large blocks of this sandstone spall off, especially the thick layers above the grey shale on which the block is resting.

Group of specimens of the smaller *Equisetum aruense.*

Calamite pith cast from the above driftwood "pocket." The genus *Calamites* is restricted to the elongate trunks of these reedy plants.

Close-up of previously shown calamite placed upon calamites external impression.

Boulder of stream deposited Warrensburg-Moberly Sandstone on the campus of Florissant Valley Community College with calamite pith-cast, internal mold (left), and *Calamites* impression (right.)

Warrensburg-Moberly channel sandstone overlying grey shale of the Mid-Pennsylvanian Marmaton Group. I-70 modifications, 2002.

Coal "smut" layer near the bottom of the previously shown blue-grey shale. Coal smut layers represent the existence of former coal swamps where either plant and plant material was thin or this material was not buried quickly enough to be preserved as a coal bed.

Calamite pith casts being collected, including the previously shown specimen.

Calamite bearing channel sandstone (Warrensburg-Moberly Sandstone), exposed during I-70 modifications, 2002. Northern St. Louis County, Missouri.

Uppermost layers of Warrensburg-Moberly channel sandstone overlain by former I-70 sub-grade and concrete pavement. Many persons traveled on this surface, which was one of the first stretches of interstate highway built during the Eisenhower Administration.

Outcrop of river-channel sandstone, southwestern Missouri, 1958. These "channel deposit" sandstones can form upland ridges, as they are more resistant to erosion than are the softer shale beds into which the sand-filled channels were deposited. This outcrop forms a high ridge at the edge of the Ozark Uplift. These channel sandstones, of fresh water origin, occur widely over Missouri, Iowa, and Illinois. Sometimes they carry pockets of "fossil driftwood," which may include *Calamites* that grew along the stream-channel banks.

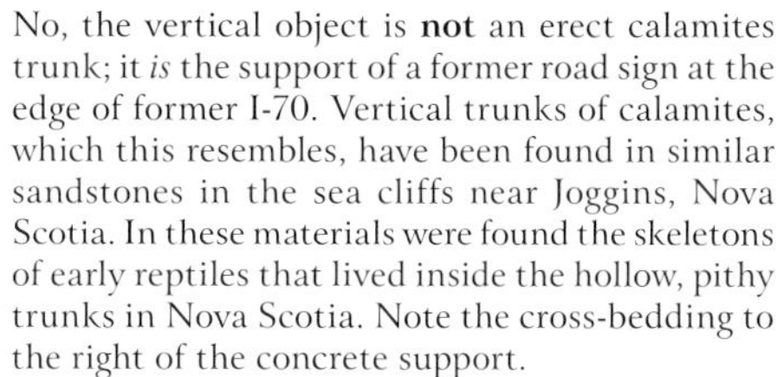

No, the vertical object is **not** an erect calamites trunk; it *is* the support of a former road sign at the edge of former I-70. Vertical trunks of calamites, which this resembles, have been found in similar sandstones in the sea cliffs near Joggins, Nova Scotia. In these materials were found the skeletons of early reptiles that lived inside the hollow, pithy trunks in Nova Scotia. Note the cross-bedding to the right of the concrete support.

Close-up of channel sandstone, southwestern Missouri.

River channel sandstone in northern Alabama formed in the same manner at the same age as those of Missouri and Illinois. Note cross-bedding in the upper layers, a characteristic of river channel deposits.

Mid-nineteenth century etching of a bluff in southern Iowa composed of this channel deposited Pennsylvanian (Upper Carboniferous) sandstone. Nineteenth century geologic reports often included such illustrations. These reports were often done to evaluate local geology and give information used to ascertain whether the examined lands were to be retained as mineral lands or were to be homesteaded as agricultural lands once statehood was granted.

Sphenophyllum sp.: Sphenophyllum was an herbaceous arthrophyte that covered the surface of Pennsylvanian coal swamps. Yellow shale of the Pleasanton Group, northern St. Louis County, Missouri. (Value range F.)

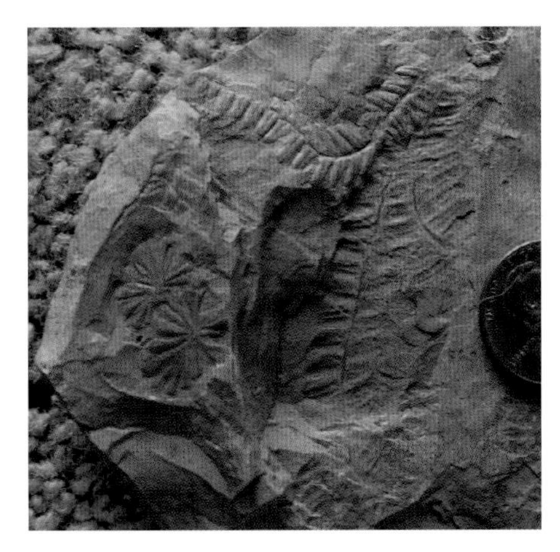

Sphenophyllum (left) with fern (*Pecopteris* sp.) from mudflat deposits associated with Mid-Pennsylvanian channel deposits in northern St. Louis County, Missouri. *Sphenophyllum* has blunter leaves than found on the foliage of calamites. (Value range G.)

Asterophyllites sp.: leaf whorls of a small variety of calamites. Individual leaves of *Asterophyllites* are more slender and elongate than are those of the herbaceous *sphenophyllum*. Its leaves also were more slender and elongate than those of *Annularia*. *Asterophyllites* was a small, herbaceous arthrophyte similar to *Equsitum* sp. Yellow shale of the Pleasanton Group, upper Mid-Pennsylvanian, northern St. Louis County, Missouri. (Value range F.)

Asterophyllites (*Annularia*) sp.: Cherokee Group, Hampton Avenue, St. Louis, Missouri. An arthrophyte specimen from an outcrop in the St. Louis area, which yielded fossil plants from the 1920s to the early '50s.

Annularia stellata: Annularia, the leaf whorls that resemble a flower, are probably the best known arthrophyte foliage of the Pennsylvanian Period. It was born on branches that extended perpendicularly from the calamites trunk. Specimen from yellow shale of the Pleasanton Group, northern St. Louis County, Missouri. (Value range F.)

Annularia stellata: part and counterpart of previously shown specimen.

Sphenophyllum, Asterophyllites, Annularia, and Calamite Reproductive Organs

Sphenophyllum was a small, herbaceous arthrophyte that grew profusely as an understory or ground cover plant in Late Carboniferous forests. Here is a gallery of these interesting and attractive plants.

Annularia radiate calamites leaves from Upper Pennsylvanian strata (Tonganoxie Sandstone) near Ottawa, Kansas. Numerous clear fossil plants have come from yellow siltstone beds near Ottawa, eastern Kansas. (Value range F.)

Annularia stellata: northern France. As is the case with other Upper Carboniferous coal swamp plants, specimens of calamites and its foliage are essentially identical to those of eastern North America. (Value range F.)

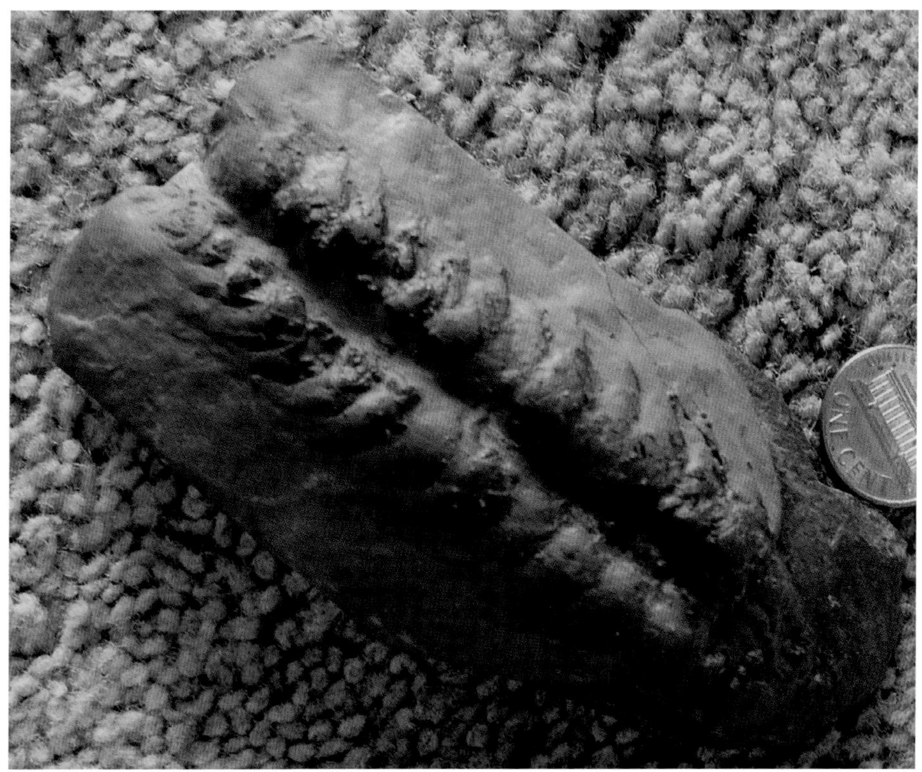

Calamostachya sp.: this is the cone-like reproductive organ of Calamites (catkin), the specimen being preserved in an ironstone concretion. It came from the Braidwood, Illinois, area, considered to be one of the world's major paleontological "windows." (Value range E.)

Equisetum sp.: small arboreal form of a living arthrophyte, which is somewhat similar to *Sphenophyllum*. Little Tonzona River, Alaska.

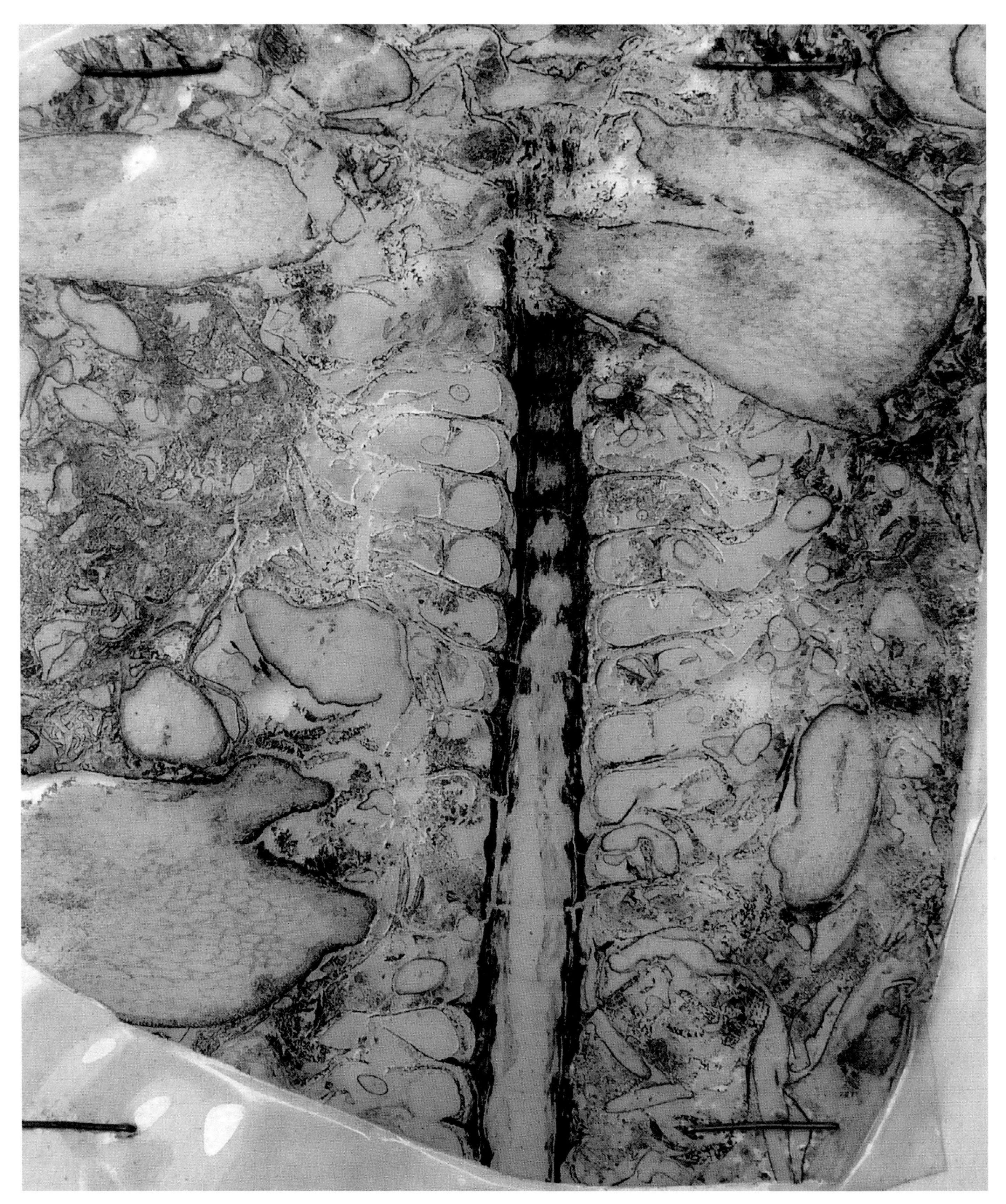

Calamostachya sp.: Calamite reproductive organ preserved in a coal ball. This is what is known as an acetate peel. It was taken from an especially nice specimen of this genus. Individual megaspores can even be seen upon close examination of the peel. These peels—taken from what are known as coal balls (limestone concretions found in coal seams that have preserved plant tissues)—offer a paleontological technique that allows for the study of ancient fossil plants in a similar manner to that used with modern plants. Coal ball from near Berryville, Illinois. *Courtesy of Henry Andrews, Washington University, St. Louis*

Sphenophyllum sp.: a group of leaf whorls preserved in grey shale of the Warrensburg-Moberly Sandstone (Pleasanton Group) from north of Shelbina, Missouri. *Sphenophyllum* was an herbaceous arthrophyte common in the understory of Pennsylvanian coal swamps. (Value range G.)

Sphenophyllum tenuifolium: this is a species of *Sphenophyllum* in which the leaf whorls are more slender than normal and somewhat suggest the leaves of Calamites. Mid-Pennsylvanian, Windsor, Henry County, Missouri.

Sphenophyllum sp.: robust specimen of this herbaceous arthrophyte, probably from an older plant than that shown above. Shales of the Pleasanton Group, Middle Pennsylvanian, Shelbina, Missouri. (Value range G.)

Group of leaf whorls of *Sphenophyllum* from near Windsor, Missouri, Henry County, Missouri.

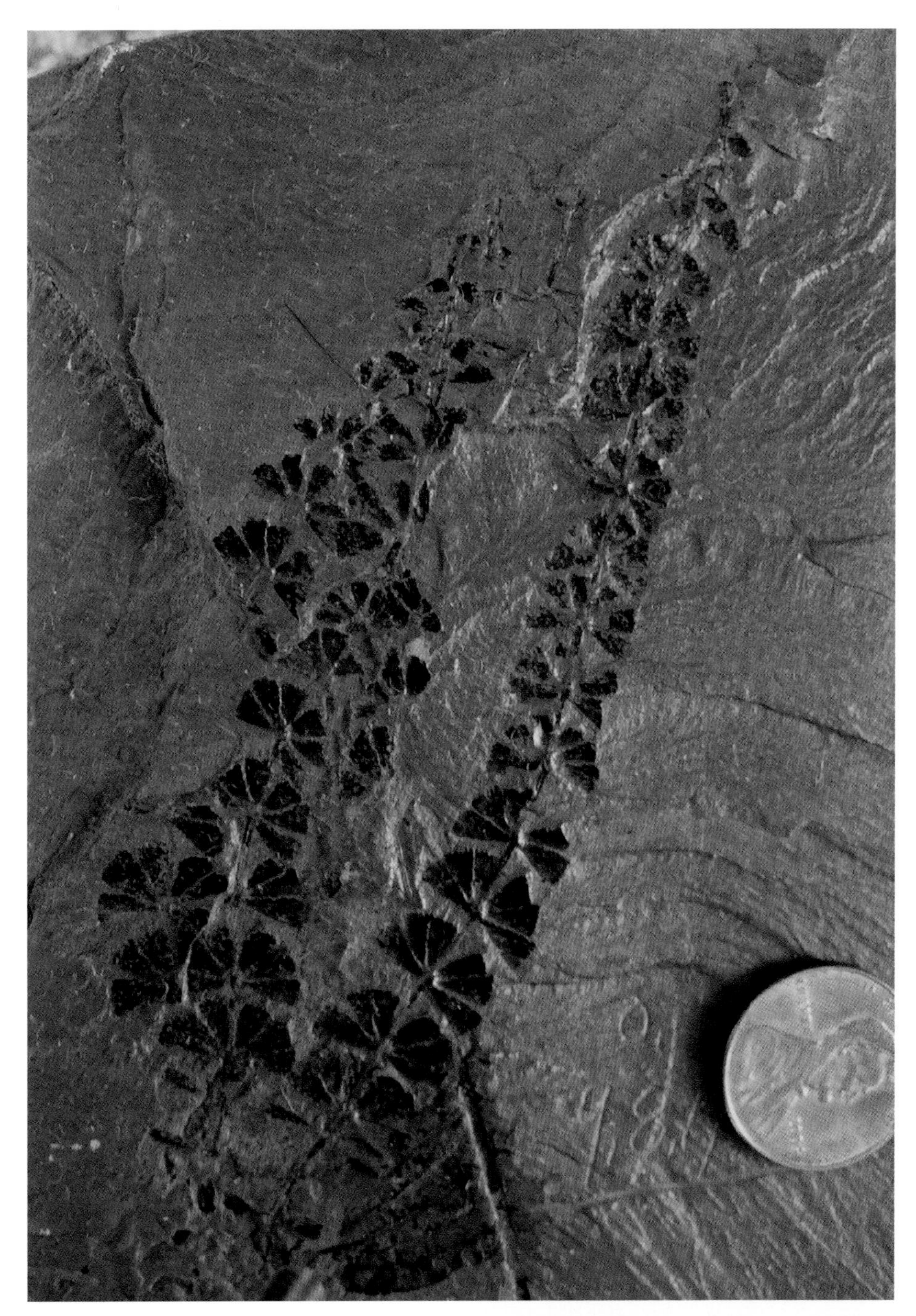

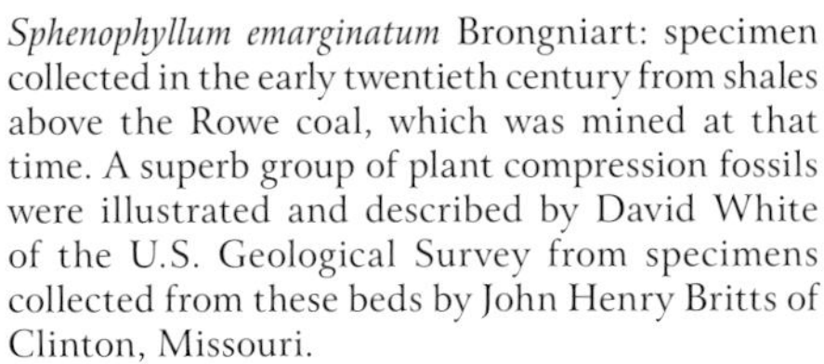
Sphenophyllum emarginatum Brongniart: specimen collected in the early twentieth century from shales above the Rowe coal, which was mined at that time. A superb group of plant compression fossils were illustrated and described by David White of the U.S. Geological Survey from specimens collected from these beds by John Henry Britts of Clinton, Missouri.

Same specimen as shown previously under different lighting. Notice the date (9/8/1905) inscribed on the specimen. These inscribed dates are found on most specimens collected by John Britts late in the nineteenth and early in the twentieth century. Britts was a Civil War veteran who, after the war, settled in western Missouri, where he practiced medicine. Like many other physicians of the nineteenth century, he maintained a serious interest in fossils and paleontology and contributed to it, often by avidly collecting specimens.

Sphenophyllum emarginatum to the right of *Neuropteris* sp. Windsor, Henry County, Missouri.

Sphenophyllum emarginatum preserved as a brown (lignitic) compression rather than the black (coal-like) compression usually seen with foliage of Paleozoic plants. Middle Pennsylvanian, Windsor, Henry County, Missouri. (Value range G.)

Sphenophyllum emarginatum Brongniart, Middle Pennsylvanian, Southern, Indiana. Pennsylvanian fossil plants, especially those of the coal swamps, are quite cosmopolitan over the eastern U.S. This specimen of *Sphenophyllum* is identical to those from western Missouri, which are of the same age.

Sphenophyllum majus (right) next to *Aleothopteris* sp. Middle Pennsylvanian, Clinton, Henry County, Missouri. J. Britts, collector. Specimen originally from the collection of St. Louis University Geology Department. Specimens collected by John Britts were distributed to geology departments all over the country, including those of St. Louis University, Washington University, Missouri School of Mines, and University of Missouri-Columbia. *Courtesy of the St. Louis Science Center*

Calamites from Deep Sea Trench or Geosyncline Deposits

During the Carboniferous (Mississippian and Pennsylvanian periods) a subsiding trench existed, at what at that time was the edge of Laurentia or North America. Numerous braided streams like those mentioned previously under "St. Louis Area Calamites" flowed into this trench. As the trench subsided, large amounts of sand, clay, and mud filled it. These sediments often containing driftwood, which included the stems and trunks of calamites. This trench, known as the Ouachita Geosyncline, eventually underwent lateral compression from tectonic forces and became the Ouachita Mountains of west central Arkansas and eastern Oklahoma. The flood of sediment that filled this trench was preceded in the Devonian Period by an influx of silica (possibly from volcanic sources) forming what is known as the Arkansas Novaculite. This is followed by the Stanley Shale, a deep sea deposit known as flysch. Often flysch is slightly metamorphosed, which gives it a silvery appearance. This thick meta-shale is followed by the Jackfork Sandstone, a thick sequence of sandstone that can yield nice specimens of what were waterlogged lepidodendrons (Chapter Five). These Mississippian age strata are overlain by Pennsylvanian strata of the Atoka Formation and the Hartshorne sandstone—the latter producing some excellent calamites, specimens that also were carried originally into the subsiding trench as waterlogged plants.

Slightly metamorphosed shale from deep sea sediments (flysch) of the Mississippian Stanley Shale of the Ouachita Mountains, Arkansas. These sediments were deposited rapidly in a subsiding basin or trough at what, in the late Paleozoic, was the edge of the North American continent. Occasionally waterlogged plants would be carried in with the accumulating mud to be preserved as fossils.

Shiny, light-reflecting hard, slaty shale of the Mississippian Stanley Shale, Hot Springs County, Ouachita Mountains, Arkansas.

Boston Mountains of the southern Ozarks, Arkansas. A thick sequence of (mostly) non-marine strata (Atoka Formation) consisting of inter-bedded shale and sandstone makes up these mountains. This rock represents the margin (or foreland facies) of a thick "wedge" of sediments deposited in what was a rapidly subsiding trench or trough that, during the late Paleozoic, was at the southern edge of Laurentia and what later would become part of the North American continent. Coal beds, trace fossils, and fossil plants, the latter of which were sometimes carried in a waterlogged condition into the subsiding trough, occur in the foreland facies of the Atoka Formation.

Archaeocalamites sp.: an early calamite from the previously shown locality in Stanley Shale, Hot Springs County, Arkansas.

Calamites from the Atoka Formation. Calamites and other stems and trunks of Pennsylvanian plants are found in stream deposited sediments as pieces of "driftwood," which were carried into the rapidly subsiding and filling trench or geosyncline occurring south of what is now the Boston Mountains. (Value range G.)

Geology students at the Mt. Magazine, Arkansas, overlook. Mt. Magazine is one of the highest points in the U.S. between the Appalachians and the Rockies. It is formed from part of a thick wedge of sediments, which was folded and thrust northward to form the Ouachita Mountains.

View from an overlook at almost 3,000 feet on Mt. Magazine, Arkansas. The Arkansas River valley is below, a valley underlain by a thick sequence of late Paleozoic sediments deposited at what, at the time, was the edge of North America. In the distance can be seen the Ouachita Mountains, a fold mountain range made up of deep sea Paleozoic sediments deposited in a deep trench (or geosyncline) and later piled up at what was at the time the continent's edge.

Calamites suckowi: Mt. Magazine, Arkansas. Nodes on this (once) waterlogged calamite section vary in their spacing. Middle Pennsylvanian, Hartshorne Sandstone. (Value range F.)

Calamites sp.: a small "inflated" stem section with distinct nodes from a mass of (once) waterlogged plants found at the foot of Mt. Magazine. Hartshorne Sandstone. (Value range F.)

Calamites from a zone of fossil plants exposed along the southern ascent road to Mt. Magazine, Arkansas. Mt. Magazine, one of the highest points in the U.S. between the Rockies and Appalachian Mountains, is composed of thick sandstone beds deposited by rivers that emptied into a subsiding trough during the middle part of the Pennsylvanian Period. The specimen at the bottom has distinct nodes and the example at the top has the nodes further apart. In some Calamite taxonomies, these would be different species. What (probably) is represented here are specimens that came from different portions of the plant or they came from plants of different maturities. (Value range F.)

Calamites with furrows between the nodes being absent or vague. Hartshorne Sandstone, Mt. Magazine, Arkansas.

A "Gallery" of Calamites and Related Genera

Most of the Calamites in this section were collected by the author as a boy in the 1950s. These were the first good fossils (along with the Mississippian coral *Lithostronella*) he collected. Their acquisition putting the "seal" on his commitment to fossils, paleontology, and geology—a fascination with science accompanied by the thrill of discovery almost literally in one's own backyard!

Part of the northern portion of the St. Louis metropolitan area is in what geologists call a "structural basin." The center of a basin contains sediments of younger geologic age than those which occur along its margins—the Calamites coming from the late Mid-Pennsylvanian age rocks at the basins center. At one time, these rocks cropped out at many places; however, most outcrops are now covered by urbanization. Most of the Calamites came from the Warrensburg-Moberly Sandstone, with most collected between 1952 and 1962. In some of the large Calamites, it was hoped to find skeletons or bones of early reptiles, the occurrence of which the author was aware of from nineteenth century geology books he had acquired as a child. He hoped to find these where large calamite trunks were discovered exposed in gullies—**but, no such luck!**

Calamites specimen with very distinct nodes and furrows. Warrensburg-Moberly Sandstone, northern St. Louis County, Missouri. (Value range F.)

Warrensburg-Moberly Sandstone exposed along Wabash Railroad tracks, 1970. Weathering has brought out what is known as cross bedding, a characteristic of river channel deposits. Some nice calamite trunks occurred in these no-longer-exposed beds. Bed number five (below) is an ancient stream deposit composed of gravel. It is younger than Pennsylvanian but is Pre-Pleistocene in age.

An ancient river-channel deposit (Warrensburg-Moberly Sandstone) in late Mid-Pennsylvanian strata exposed during construction of I-70 in northern St. Louis County, Missouri, 2003. This river channel deposit was first collected in the 1950s by the author, who discovered how the plant masses occurred. When rerouting of I-70 took place in 2002, these zones were exposed. The black layer at the top is the old shoulder of I-70.

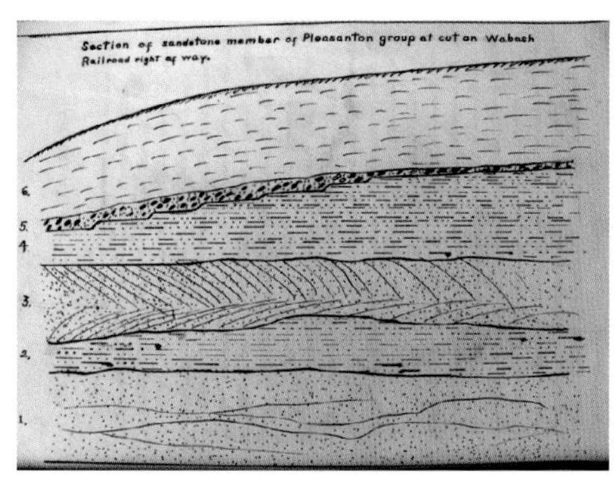

Diagram of the above strata exposed in former Wabash Railroad cut.

Terminal portion of Calamites, probably from a relatively young plant. (Value range E.)

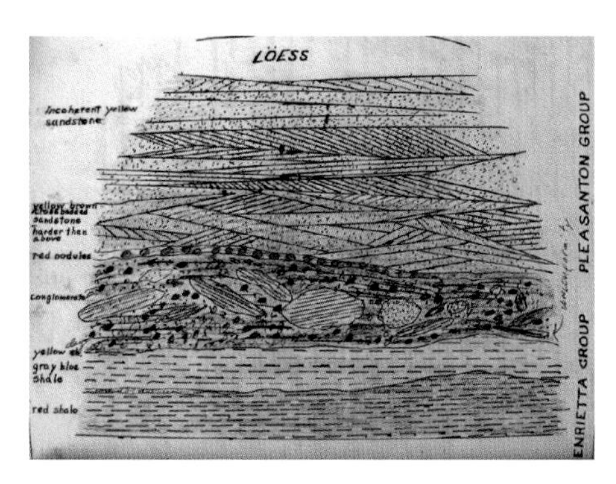

Diagram of Calamite bearing strata just below that of the above two sequences of strata. Loess shown at the top of the diagram is of Pleistocene (ice age) time.

Typical pith cast of *Calamites*. Here the nodes are close together, probably this is from the lower (and older) portion of the plant. (Value range F.)

Long Calamite pith-cast from the previously shown locality in northern St. Louis County, Missouri, collected in 1954.

Calamites pith cast with some mineralized, secondary wood. Here the nodes are spaced the same distance as the pith cast diameter. (Value range E.)

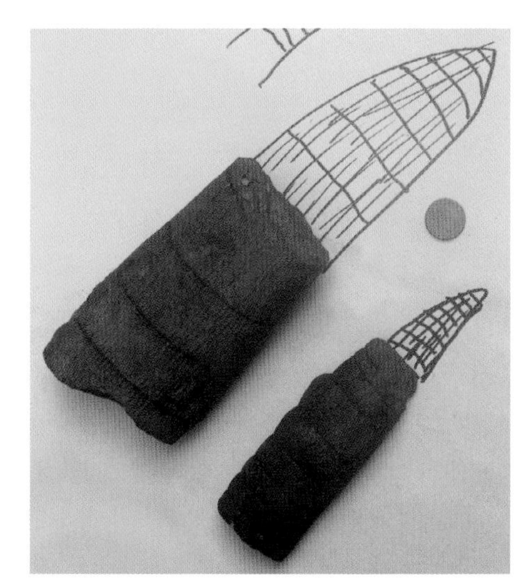

Calamite sections from near the plant's terminal portion. (Value range G.)

Flattened pith of Calamites.

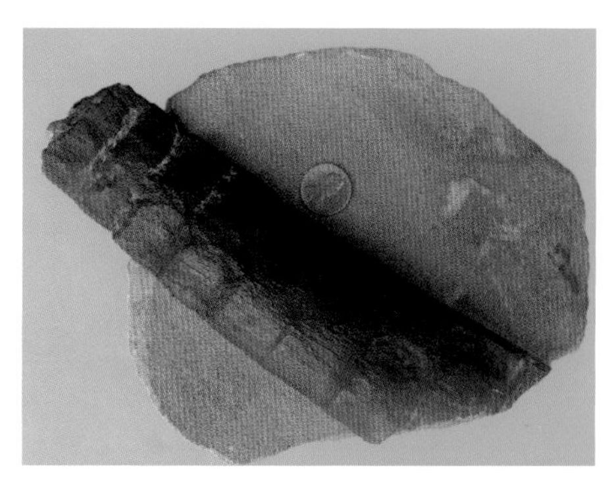

Closely spaced nodes and vague longitudinal furrows characterize this Calamite specimen. (Value range F.)

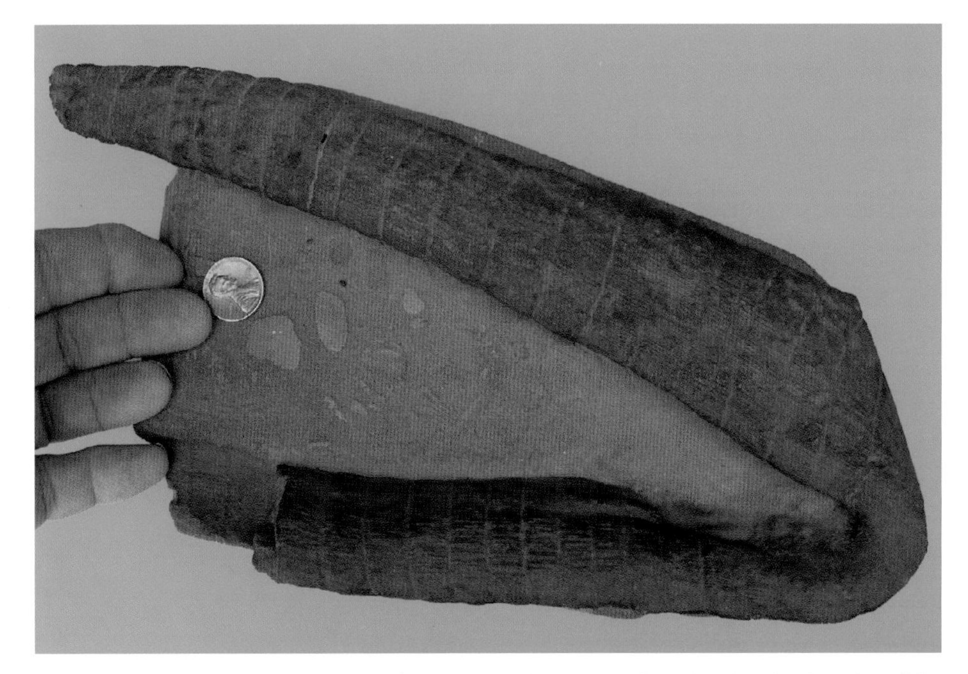

What is probably the terminal portion of a calamite pith cast. Note how the plant has been bent. The reedy nature of Calamites, especially the terminal portion, allowed for this. From the Wabash Railroad cutting, Normandy, Missouri. (Value range E.)

Calamite pith cast flattened at the left but still round at the right. Reedy construction of these plants allowed them to readily become flattened. (Value range E.)

Calamodendron or *Bornia* sp.: what might at first be confused is this pith cast. It was considered in the late nineteenth century to be a type of gymnosperm or conifer. Its variable width and segments is suggestive of one of the other widespread Pennsylvanian plants, Cordaites. Cordaites are believed to be a possible ancestor to the conifers. It has closely-spaced nodes suggestive of calamites, but lacks the longitudinal furrows. This fossil was associated with the previously shown driftwood mass—a mass consisting of mostly Calamites stems and trunks, most poorly preserved. (Value range F.)

Calamites sp.: typical pith cast, Middle Pennsylvanian, southern Indiana.

Calamites sp.: pith cast, Joggins, Nova Scotia. This came from thick Pennsylvanian sandstones that contain calamites trunks in which some of the oldest reptiles known were found.

Joggins, Nova Scotia, Calamites

A thick sequence of Pennsylvanian strata crops out along the Bay of Fundy in Nova Scotia, Canada. Here in the mid-nineteenth century, upright trunks of Calamites were found containing the bones of small reptiles, which apparently used the hollow trunks for protection. This early occurrence of reptiles is still unique and has recently been made a UN World Heritage Site.

Outcrop of dipping Lower Carboniferous (Pennsylvanian) strata at Joggins, Nova Scotia, Canada.

Canadian postage stamp with early reptile reconstruction of *Hylonomus lyelli*. Skeletons of these early reptiles were found by J. William Dawson in the interior of large calamite trunks at Joggins, Nova Scotia, in the 1840s, where they were some of the first Paleozoic reptiles to be found anywhere. Vertebrate fossils of Pennsylvanian age, especially amphibians and reptiles, are very rare, especially those associated with coal swamps, as was *Hylonomus*.

Calamites and Filled Sink (Paleokarst) Deposits, Central Ozarks

Sinkholes developed in Lower Paleozoic dolomite and filled with Pennsylvanian age sediment occur in the central and northern part of the Ozarks of Missouri, many of which previously were mined for refractory (fire) clay. While usually lacking fossils, occasionally fossil plants were found in the fire clay, some of which were unusual. Calamites are among of the more common examples found.

Slender calamites stem from Pennsylvanian filled sink deposits near Rolla, Missouri. (Value range G.)

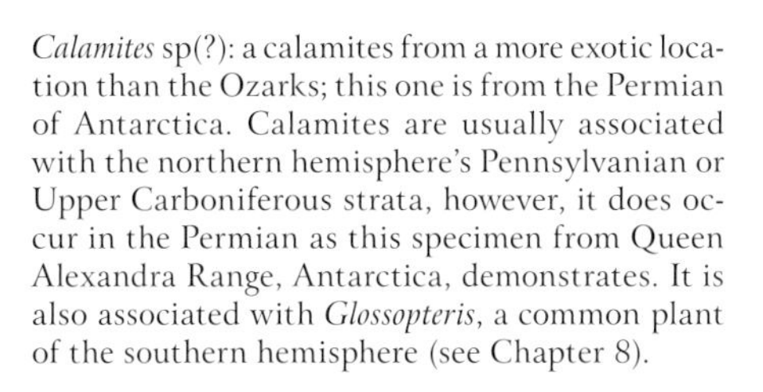

Calamites sp(?): a calamites from a more exotic location than the Ozarks; this one is from the Permian of Antarctica. Calamites are usually associated with the northern hemisphere's Pennsylvanian or Upper Carboniferous strata, however, it does occur in the Permian as this specimen from Queen Alexandra Range, Antarctica, demonstrates. It is also associated with *Glossopteris*, a common plant of the southern hemisphere (see Chapter 8).

Ancient sinkhole (paleokarst) filled with Pennsylvanian sediment. This exposure, along former U.S. Highway 66 near Rolla, Missouri, shows thin bedded Pennsylvanian strata consisting of shale, fireclay, and sandstone. These beds fill a solution cavity (sinkhole) developed in Lower Ordovician Dolomite. Such "filled sinks" occasionally contain sediments (shale and fireclay) that preserve Pennsylvanian plants, some of which were different from the usual "coal swamp" flora. The large rounded rock was thought to be a large concretion; however, when it fell from the cut in the 1970s it was found to be an iron-oxide-clad rounded mass of the Ordovician dolomite.

Arthrophytes in the "Mazon Creek" Fossil Beds, Braidwood Illinois Area

A variety of well preserved arthrophytes occur in the famous Mazon Creek ironstone nodules, which occur southwest of Chicago, Illinois. These occur in reddish ironstone concretions that weather out from shale beds exposed during the mining of coal in the 1930s through the '50s. Large numbers of these fossil bearing concretions were collected, which frequently are seen at rock, mineral, and fossil fairs.

Large numbers of the plant bearing ironstone concretions have been distributed among fossil collectors, collector-dealers, and dealers. These often are available at low prices at mineral-fossil-rock shows where (in 2011) a flat of them could be had for $25.00.

"Spoil piles" in the Braidwood-Coal City area, 1953. This is part of the plant-fossil bearing region known as the Mazon Creek fossil beds. These still produce fossiliferous ironstone concretions, but not in the quantity they did decades past.

Astrophyllites equisetiformis: these leaves from a Calamites plant have a star-like leaf whorl made up of slender leaves. The specimen was probably collected recently as it has accumulated a stain of gypsum, which can be difficult to remove. Specimens collected in decades past usually lack this whitish material. (Value range F.)

Astrophyllites equisetiformis: specimen from the Braidwood, Illinois, area displayed at a North American Paleontological Convention, Field Museum, Chicago, 1978.

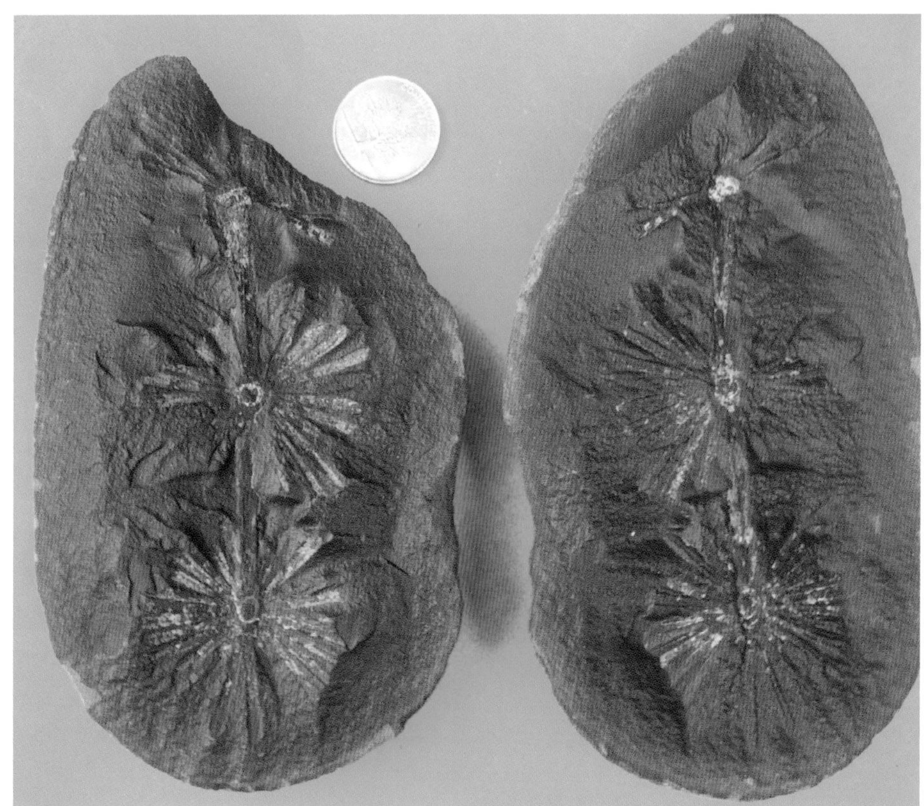

Annularia stellata: probably the most common species of *Annularia* in the Braidwood, Illinois, area. (Value range F.)

Annularia sp.: some Braidwood ironstone concretions are grey rather than reddish. These were recently uncovered and were previously deeply buried in the spoil piles, which today are sometimes being leveled for housing developments. *Courtesy of John Stade*

Annularia stellata, single whorl: the most common form of *Annularia* from Braidwood and the Mazon Creek fossil beds. (Value range G.)

Annularia radiata: single whorl in an ironstone concretion. This comes from near Terre Haute, Indiana, where a flora (and fauna) in ironstone concretions similar to those of Braidwood, Illinois, occur. (Value range G.)

Annularia stellata: Braidwood, Illinois, area. (Value range G.)

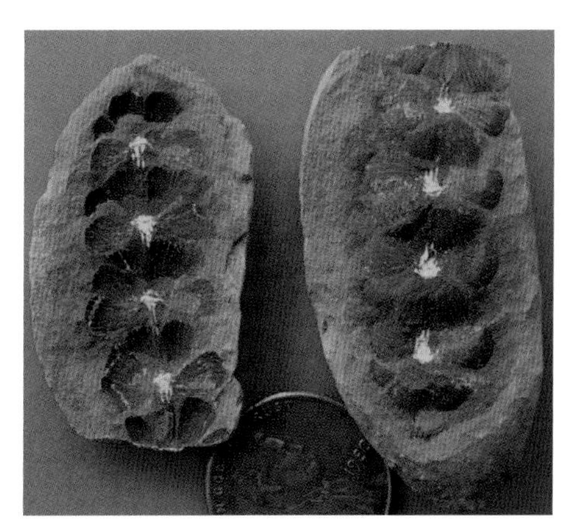

Sphenophyllum sp.: Braidwood, Illinois.

Sphenophyllum emarginatum Brongniart: an herbaceous arthrophyte that grew in the understory of the coal swamps. Braidwood ironstone concretion. (Value range F.)

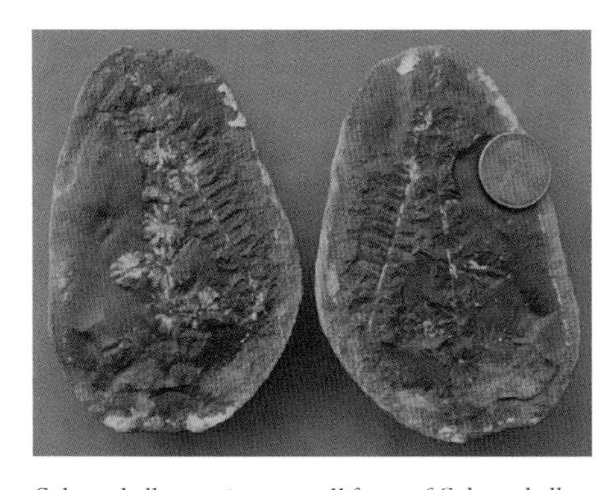

Sphenophyllum majus: a small form of *Sphenophyllum* associated with *Pecopteris* sp. (true fern impression).

Calamites ramosus in ironstone concretion from Braidwood, Illinois. (Value range H.)

Calamites sp.: a form of Calamites in which the distance between nodes (internodes) are considerably longer than is the stem width.

Calamites in Coal Balls and Coal Ball Peels

Coal beds, when covered by marine deposits (as they commonly are in the U.S. Midwest), can have some of their plant material infiltrated and sealed with calcium carbonate, forming what are known as "coal balls." Exposed either in mining or when a stream cuts into a coal seam, these hard concretions can be a paleobotanical bonanza. When sliced with a rock saw, they often will contain well preserved plants. Such plants are preserved in a way that allows them to be studied in the same manner as modern plants. Various types of cells are preserved in them and fine detail of plant anatomy is revealed. Sometimes large numbers of coal balls have to be sectioned to find fossils like those shown here—especially ones like the calamite catkin shown on page 106. When a significant coal ball fossil is found by sawing, it is smoothed and then etched with dilute acid. It is then saturated with acetone and covered by a sheet of clear acetate film. When the acetate evaporates, the film can be removed along with embedded material from the coal ball. In this way, a small portion of the plant material is transferred to the acetate sheet, forming what is known as a "peel." Peels resemble negatives of photographic film. The following arthrophyte fossils are images from such "peels." The acetate film technique using coal ball peels in paleobotany was pioneered by paleobotanist Henry N. Andrews of Washington University in the 1950s and '60s. The author took a number of courses in the 1960s from Dr. Andrews, from whom he obtained a number of coal balls and their peels.

Arthropithys sp.: crushed Calamite trunk with secondary wood in which the pith has been flattened. A large amount of secondary wood is present (the criteria for the genus *Arthropithys*). Coal ball from Berryville, Illinois.

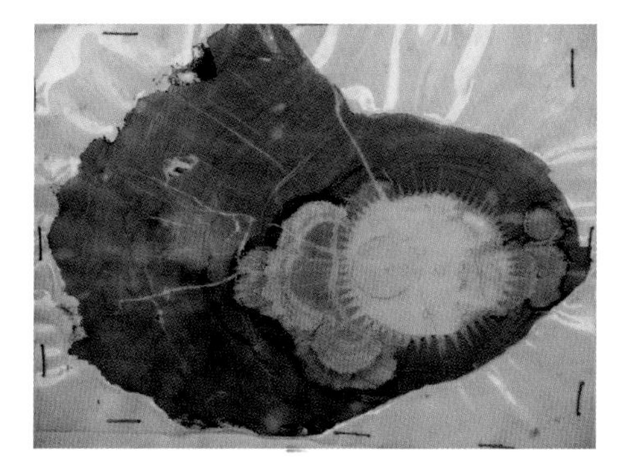

Arthropithys sp.: large Calamite trunk with thick secondary wood (over 10 cm [4"]) surrounding the pith, which can be seen here as a white, radiating pattern. The rays are the longitudinal furrows of the pith cast. Coal ball peel from Berryville, Illinois. *Courtesy of Henry N. Andrews*

Petrified calamite with a considerable amount of secondary wood (*Arthropithys* sp.) from Middle Pennsylvanian coal ball, Berryville, Illinois.

Glossary

Coal balls: calcareous concretions that form in the plant material of a future coal seam. The embedding of plant tissue by calcium carbonate takes place **prior** to conversion of the plant material to coal so that fine structure of the plant is present, a phenomena generally **not found in coal itself**.

Coalification: the conversion of plant material by both time and pressure into coal. In the coalification process, any tissue-structure of plants generally is lost.

Flysch: sediments deposited rapidly in a subsiding basin or a deep sea trench. Sometimes these sediments on their bedding surfaces, have trace fossils produced by deep sea organisms. Named for deep sea sediments that now make up portions of the Alps of Europe.

Foreland facies of a clastic wedge: in reference to a subsiding trough or geosyncline, this is the thinnest portion of the thick wedge-shaped mass of sediments deposited, usually under rather rapid conditions (geologically speaking).

Geosyncline: a trench or subsiding basin that can be filled with a great thickness of sediment. This sediment may be cooked or metamorphosed by both deep burial and tectonic forces, which follow formation of a geosyncline. Such thick sequences of sediment also can become the site of fold mountain ranges. Both the Appalachians and the Ouachitas are sites of former geosynclines. The geosynclinal theory was first proposed by James Hall, a nineteenth century paleontologist-geologist. It is now incorporated into Plate Tectonics but the concept is still a valid one.

Laurentia: that portion of Laurasia that (when it split up) became North America.

Subsiding Trough: a portion of the earth's crust that subsides and, in the process, accumulates sediments. Such a trough can be a deep sea trench or it can be what in geology is known as a geosyncline. The Ouachita Mountains of Arkansas and Oklahoma were formed from a rapidly subsiding trough that accumulated a thick sequence of sediment during the Mississippian and Pennsylvanian periods.

Chapter Seven

Primitive Conifers

Cordaites

Cordaites were late Paleozoic plants with distinctive, strap-like leaves that resemble those of an iris. They are found associated with both the coal swamp flora as well as with more upland environments. Cordaites are considered by many paleobotanists as primitive conifers.

Tree-sized cordaites of the Permian Period are on the right. *Courtesy of Dorothy J. Echols*

Cordaites Stems and Trunks

The woody portion of the cordaites plant, like most Paleozoic plants, was pithy and punky. Pith casts and leaves are the most frequently seen fossils. Cordaites pith casts often exhibit what resembles somewhat irregular segmentation. These pith casts are given the form genus *Artisia*.

Artisia sp.: distinctive pith casts of cordaites. Like most Paleozoic plants, the trunk and branches of cordaites was pithy and reedy. These are sandstone fillings of the interior of cordaites stems. The segmented appearance of the plant's interior is characteristic of what is believed to be an early conifer. Atoka Formation, Lower Pennsylvanian (foreland facies), Boston Mountains, Jasper, Arkansas. (Value range F.)

Artisia sp.: cordaites pith casts from Middle Pennsylvanian Warrensburg-Moberly channel sandstone. Randolph County, Missouri. (Value range F.)

Flattened cordaites pith cast that came from the Lower Pennsylvanian, Atoka Formation, Jasper, Arkansas. (Value range G.)

Artisia sp.: cordaites pith cast that has been removed from an ironstone concretion, Braidwood, Illinois. (Value range G.)

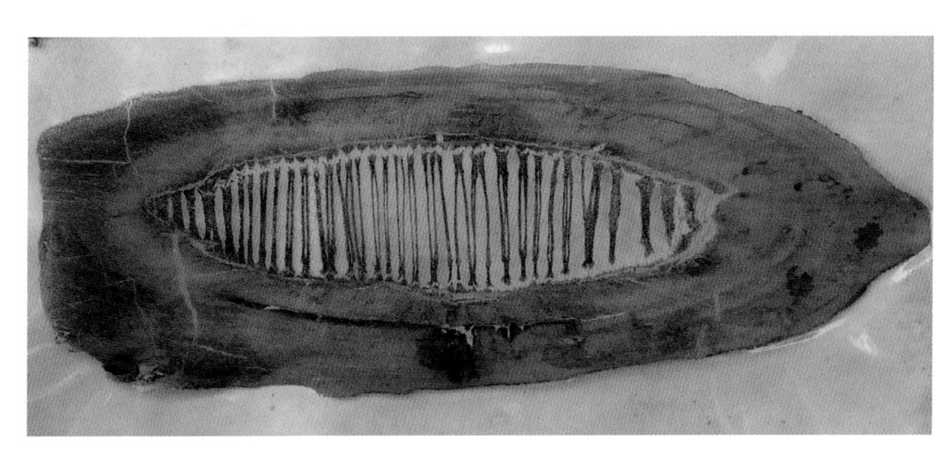

Mesoxylon sp.: transverse section (acetate peel) of cordaites showing its distinctive interior partitions preserved in a coal ball. Middle Pennsylvanian, Berryville, Illinois. *Courtesy of Henry N. Andrews*

Cordaites Leaves

The leaves of cordaites were strap-like, resembling the leaves of an iris or of corn. The large, tough, strap-like parallel veined leaves survived transport well. They are usually easily recognized, even as fragments. Clusters of small cordaites-like leaves are believed to originally have held the plant's spores and was the initial structure in the formation of the cone—the reproductive structure so characteristic of conifers. These "proto-cones" sometimes found associated with cordaites leaves are given the form genus of *Cordaianthus*.

Cordaites principalis: a portion of a large, strap-like cordate leaf. Some of the strap-like leaves of cordaites were well over a meter in length. The prominent one shown here was probably of that length. The parallel grooves are also characteristic of well preserved and larger specimens. Late Pennsylvanian, Douglas Group, Miami County, Kansas. *Courtesy of Steve Holley*

Cordaites sp.: strap-like cordaites leaf in sandstone. Cordaites leaves were probably similar in appearance to the leaves of the iris, although the two plants are not at all related. This is a single cordaite leaf in sandstone. Often numerous leaves of this plant are found together. Warrensburg-Moberly Sandstone, Warrensburg, Missouri.

Cordaite leaf, part and counterpart of strap-like cordaite leaf. Probably from Warrensburg-Moberly Sandstone, Warrensburg, Missouri.

Small cordaite leaf in Lower Pennsylvanian shale. Hickory County, Missouri. (Value range G.)

Another typical group of cordaites leaves from Pleasanton Group strata.

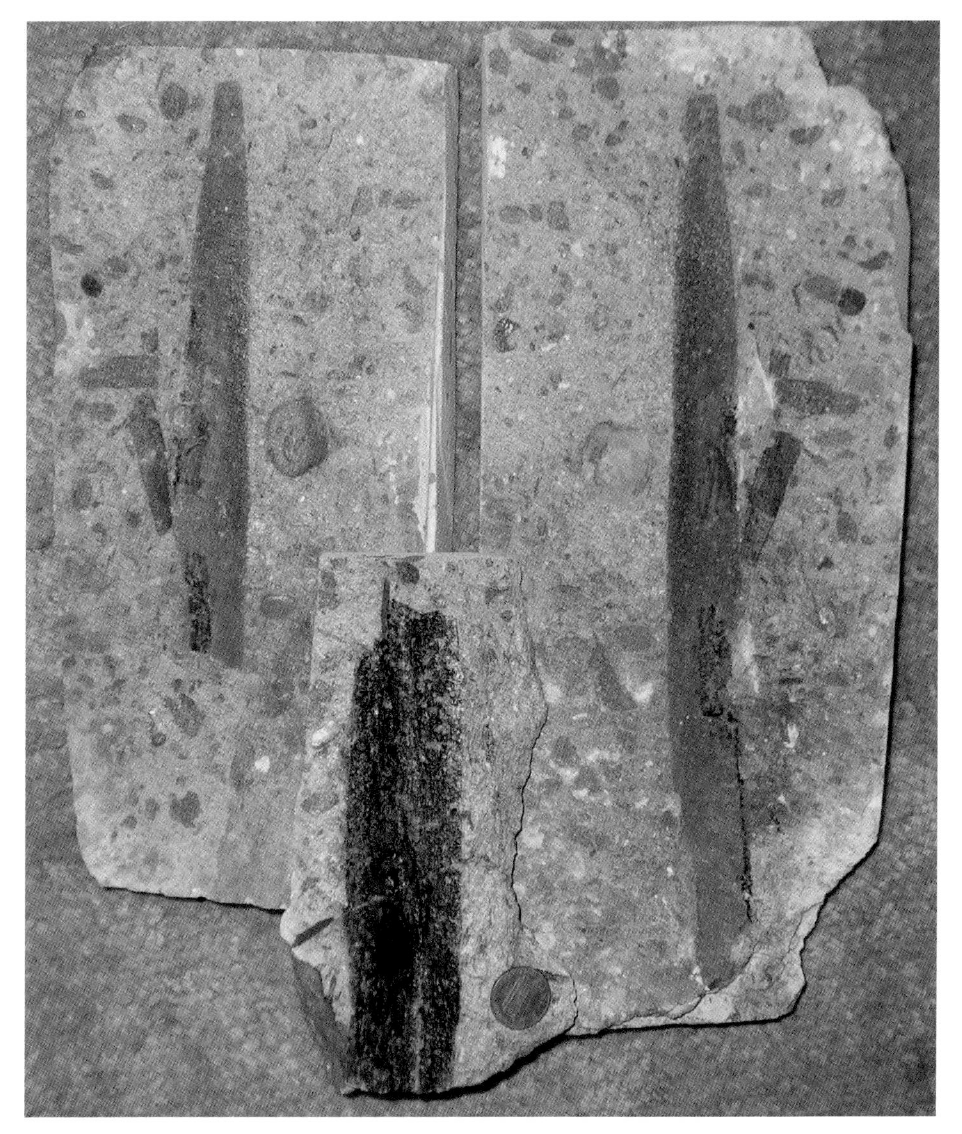

Cordaites leaves: same specimen as in previous photo. Probably from Warrensburg-Moberly Sandstone, Warrensburg, Missouri. (Value range F.)

Filled sink or paleokarst near Rolla, Missouri, 1960. This ancient sinkhole became filled with Pennsylvanian sediments and was exposed along highway 66 (later I-44) in the 1950s and '60s. The sinkhole itself was developed in Lower Ordovician dolomite, which can be seen at the left. Black layers in the white Pennsylvanian sediments contain carbonaceous material that includes macerated plant remains. This cut has now been cemented over.

Cordaites leaf in fire (refractory) clay from a sinkhole deposit of the northern Ozarks near Rolla, Missouri. Filled sinks (or paleokarsts) of the northern Ozarks were extensively mined for refractory clay from the 1930s to the 1980s. Occasionally, fossil plants were found associated with these deposits. The plants found often were of a more upland (and drier) environment compared with those plants associated with coal swamps. Cordaites were one of the most frequently occurring plants in these paleokarsts as they apparently favored a drier environment.

A group of strip-like cordaites leaves. Often numerous leaves of the cordaite plant are found clumped together like this. They are not especially attractive fossils, so they are not collected as frequently as are other common late Paleozoic plant foliage like that of Calamites and tree ferns. Upper Middle Pennsylvanian (Pleasanton Group) sediments, north St. Louis County, Missouri. (Value range H.)

Mississippian Cordaites(?)

Cordaites-like plants occur in limestone of the Middle Mississippian (Meramecian) Salem Formation in the St. Louis area and elsewhere in Illinois. They occur associated with brachiopods and bryozoans in a normal marine fauna of the mid-Mississippian and may represent plants that were growing along nearby shore areas and washed into shallow seas. Land plants in marine limestone are relatively rare as they do not usually get carried out far enough from land to become buried in the environment where limestone forms.

Cordaites sp.: Salem Limestone, Middle Mississippian. A pith cast composed of limestone from what is believed to be an early cordaites plant. St. Louis County, Missouri. (Value range F.)

Cordaites specimen in marine limestone. Note the brachiopod at the bottom. Salem Limestone, Middle Mississippian, St. Louis County, Missouri.

Cordaites sp.: carbonized wood occurs on either side of the pith cast. Vague segmentation is suggestive of cordaites with these puzzling plants. Salem Limestone.

Group of Mississippian cordaites pith casts. Salem Formation, Middle Mississippian, St. Louis County, Missouri.

Walchia (or Lebachia)

Walchia (or *Lebachia*) represents a group of late Paleozoic plants that, like cordaites, are believed to be early conifers. Whereas cordaites was a large tree, *Walchia* and related genera were small, herbaceous plants.

Walchia sp.: a branch of this early conifer preserved in non-marine (freshwater) limestone. Topeka Formation, Upper Pennsylvanian, Greenwood County, Kansas, eastern Kansas.

Walchia sp.: this plant appears first in the late Pennsylvanian, becoming more common in rocks of the Permian Period. Like cordaites, it is considered to be an early conifer.

Walchia sp. or *Lebachia* sp.: this Permian plant can be packed together in large numbers in thin layers of limy shale. When this occurs, however, preservation is generally poor. Early Permian, Abilene, Kansas.

Early Mesozoic Conifers

The Triassic, the earliest period of the Mesozoic Era, ushered in an abundance of gymnospermous plants. Whereas plants like cordaites and *Walchia* probably were early conifers, they show no undoubted relationships with them. Triassic and other Mesozoic gymnosperms, by contrast, with both compression fossils and permineralized (petrified) wood, show that they undoubtedly are fossil conifers. With the vegetation of Mesozoic conifers, there is no doubt as to the placement of compression fossils. *Taxodium* (Cypress), *Metasequoia,* and *Sequoia* are all represented—trees that are still living today.

Walchia(?): a Permian mystery fossil. It suggests *Walchia* but its leaves are rounded. Otherwise, those leaves are arranged in the manner characteristic of *Walchia*. Upper Permian, Wellington Formation, Perry, Noble County, Oklahoma.

Conifer(?): a Triassic mystery fossil. Vegetation like this can occur very early in the Mesozoic Era. This plant(?) comes from the lowermost Triassic Moenkopi Formation of northern Arizona. It also suggests a small arthrophyte (Chapter Six). The Moenkopi Formation occurs just above the late Permian Kaibab Limestone and forms the very bottom of strata representative of the Mesozoic Era in the southwestern U.S.

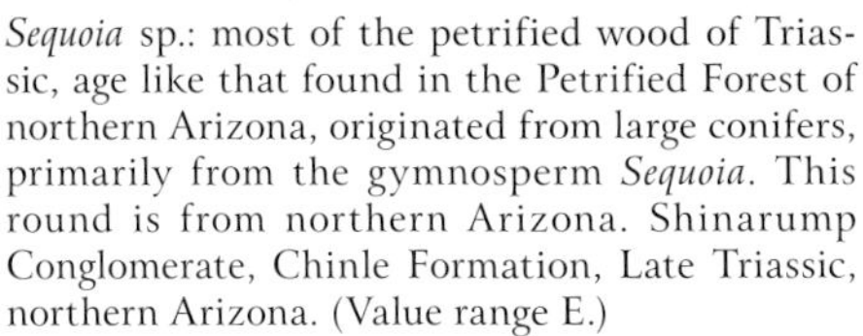

Sequoia sp.: most of the petrified wood of Triassic, age like that found in the Petrified Forest of northern Arizona, originated from large conifers, primarily from the gymnosperm *Sequoia*. This round is from northern Arizona. Shinarump Conglomerate, Chinle Formation, Late Triassic, northern Arizona. (Value range E.)

Sequoia sp.: petrified wood almost identical to that of northern Arizona is found in Madagascar. Numerous polished rounds of this early conifer wood have come through the Tucson, Arizona, show. Triassic of Madagascar. (Value range F.)

Metasequoia sp.: like cypress, Metasequoia is a recognizable conifer that goes back to the Mesozoic Era. Cretaceous, Smithers, British Columbia.

Araucaria sp.: Jurassic conifer cone—the quintessential conifer fossil. These exquisite silicified cones have come in quantity from Patagonia, Argentina. Cones like this did not exist on Paleozoic conifers—the Mesozoic heralding the existence of what are generally thought of as conifers like sequoia, metasequoia, and pines. (Value range D.)

Taxodium sp. (Cypress): this conifer has a long geologic record that, like other conifers with this type of vegetation, started in the Mesozoic Era.

Metasequoia sp.: specimen from an ironstone concretion found just above a coal (lignite) bed in North Dakota.

Chapter Eight
Glossopteris, Puzzling "Plant" Fossils, and Paleobotanical "Musings"
Late Paleozoic Red Beds

In many parts of the globe, Late Paleozoic strata (especially that of the Permian) is **red** in color. The red originates from the presence of ferric iron, which is iron oxide that has become totally oxidized by conditions that existed at the time the sediments were deposited. Those conditions involved at least partial aridity. The following fossil plants are representative of these conditions.

Pennsylvanian red beds (Fountain Formation), Garden of the Gods, Colorado Springs, Colorado, 1950.

Late Paleozoic strata (Pennsylvanian and Permian) in the Grand Canyon. Much of the upper part of the canyon consists of late Paleozoic red beds, especially Pennsylvanian (Supai Sandstone) and Permian strata.

Early conifer(?) from red beds: a more or less continuous sequence of red beds extends over the southwestern U.S. This fossil plant comes from such a sequence in northern Arizona. It occurs just above the Permian-Triassic boundary in the Triassic Moenkopi Formation, hence it is not Paleozoic in age, but it's pretty close.

Permian strata (red beds) exposed along the San Juan River, southern Utah, 1998.

Glossopteris

Glossopteris is a puzzling plant fossil found in Permian strata of the southern hemisphere. The name comes from the tongue-shaped configuration of the plants leaf, that portion of the fossil most commonly seen. Knowledge of the various species of Glossopteris is primarily obtained from compression fossils of its leaves. It has been considered variously as being a type of seed fern (pteridospermophyta), a relative to the northern hemisphere's cordaites, or even as some early representative of the angiosperms (the common seed bearing plants of today). Fossil leaves of Glossopteris are known from India (where they were first described), South Africa, South America, Australia, and Antarctica. The concept of continental drift utilized the southern hemisphere's exclusive occurrence of Glossopteris as one of its strongest paleontological supports. Most Glossopteris fossils seen on the collector's market come from Australia's state of New South Wales, where they are found as ferruginous (iron-bearing) leaf impressions in attractive hard pink and red siltstone.

Glossopteris brownii Brongniart: *Glossopteris* is a puzzling plant found in Permian strata of the southern hemisphere. It has been considered as a type of seed fern (pteridospermophyta), or perhaps a relative to the cordaites or possibly even a representative of an early group of angiosperms. Its an odd plant characterized by odd reproductive structures. Upper Permian, Duneedoo, New South Wales, Australia.

Glossopteris sp.: this leaf is lacking iron oxide, while the surrounding rocks contain it. This makes a nice contrast opposite to that of the previous specimen from Australia.

Glossopteris sp.: a "species" with a strongly bifurcating leaf. Glossopteris consists of a variety of different species. Whether these actually represent different species of the genus or are variations in leaf shape on the same plant is not known. Variations like this suggest leaf variations on the same plant, especially as the two leaf types are found together, as they are here. Duneedoo, New South Wales, Australia.

Slab with normal *Glossopteris* (center) and bifurcated form (left). Late Permian, Duneedoo, New South Wales, Australia. (Value range F.)

Glossopteris sp.: these *Glossopteris* leaves have absorbed an iron compound (ferric iron) and are preserved as an imprint formed from absorption of this red mineral. (Value range F.)

Glossopteris browniana Brongniart: the generic name *Glossopteris* comes from glosso = tongue, and refers to the plant's distinctive tongue-like shape. (Value range F.)

Glossopteris browniana: the most frequently seen species of *Glossopteris*. Most Glossopteris come from Australia, although it is also found in South America, India, South Africa, and Antarctica.

Tree Fern Trunk's Sections From the Permian of Australia

Attractive silicified trunk sections of what are believed to be tree ferns occur associated with Permian coal bearing strata in eastern Australia in both Queensland and New South Wales. Some of these trunks somewhat suggest silicified cycads of the Mesozoic Era.

Palaeosmunda sp.: silicified portions of trunk of Permian fern-like plants. Blackwater Coal Measures, Emerald, Queensland, Australia.

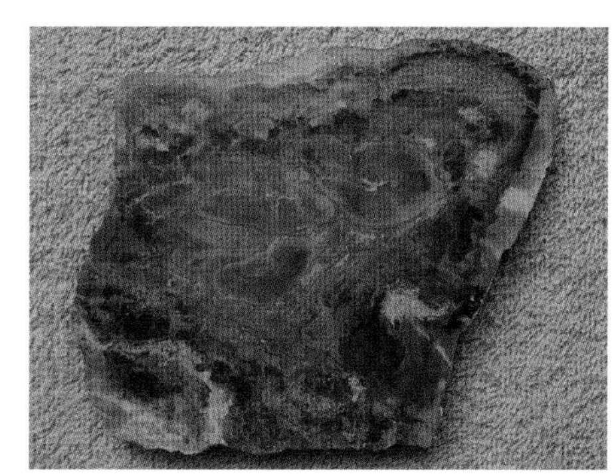

Pentoxylon sp.: a silicified tree fern trunk showing five sections (hence the name). Permian, Queensland, Australia.

Palaeosmunda willamsi: polished section of a trunk of what is believed to be a Permian tree fern. Morambah area, Queensland, Australia. (Value range E.)

Problematic Paleozoic "Plant" Fossils

Nineteenth and early twentieth century paleontological investigations and the literature on them included many descriptions of what were believed at the time to be fossil plants, especially plants collectively referred to as seaweeds. Abundant along the shores and beaches today, these often soft and gelatinous plants were logically considered to also have been present in the geologic past and to have been responsible for a variety of fossils referred to at the time as **fucoids**. This assumption came from their (often vague) stem-like and bifurcating patterns, which do resemble seaweeds, especially members of the brown algae. These seaweed-like patterns are often found in sedimentary rocks, especially those deposited under marine conditions, which includes much Paleozoic strata. Most of these "fossil seaweeds" or "fucoids" are known today as "trace fossils" and therefore were **not**, for the most part, the **sediment-filled "ghosts" of large, marine plants**. Trace fossils are the tracks, burrows, and trails of various (usually marine) organisms made from the organisms interacting with sea floor sediments. This is not to say that some of these fossils were not the actual vague impressions of seaweeds; however, it is often difficult to separate what is a trace fossil from an impression or sediment filling of a true fucoid or seaweed.

Asterophycus

One of the best known proverbial Paleozoic "fucoids," which actually is a trace fossil (a feeding trail), is *Asterophycus*. Described by nineteenth century paleobotanist Leo Lesquereux in 1879, *Asterophycus* is usually found in sandstone beds, which often occur associated with coal seams. Because of its frequent association with coal, *Asterophycus* was logically considered by Lesquereux to be some type of (probably freshwater) plant. Hence the name astero=star and its ending **phycus = plant**. Lesquereux's naming of what would later turn out to be trace fossils established a tradition in paleontology of naming mysterious and puzzling fossils as fossil plants. Other frequently found Paleozoic trace fossil names that also end in ***phycus*** include *Rusophycus*, *Arthrophycus*, and *Spirophycus*. Another puzzling fossil considered by some workers as a plant was *Conostichus*. *Conostichus* was considered as a freshwater sponge by Lesquereux—not as a plant—but later workers often considered it as such. More recently, the Treatise on Invertebrate Paleontology consider *Conostichus* as being a type of medusoid or the impression of a type of jellyfish.

Asterophycus sp. Lesquereux: a star-like trace fossil, originally described by the nineteenth century pioneer paleobotanist Leo Lesquereux. He interpreted these as a type of fossil marine plant. Lesquereux described a number of what are now understood to be trace fossils as various types of what he believed to be marine algae, viz. seaweeds. Lest one be critical of this, consider that he was the ***first*** (in the U.S.) to study these often-puzzling fossils. Scientific pioneers, like any pioneer, can make mistakes. Knowledge, after all, is a cumulative thing, building upon itself (hindsight is easier than foresight)!

Phycodes sp.: (Fucoids), Atoka Formation, Lower Pennsylvanian, Boston Mountains, Arkansas. (Value range F, an especially nice specimen.)

Phycodes sp.: another variant on this trace fossil, which was originally considered and described as a plant. Atoka Formation, Boston Mountains, Arkansas. (Value range G.)

Arthrophycus sp.: Hall, 1862. This trace fossil, originally described as a plant ("segmented" plant), was one of a large number of Paleozoic fossils described by nineteenth century paleontologist James Hall.

Chondrites sp.: these "plant-like" burrows originally were considered to be a type of brown algae (fucoids) by a number of nineteenth and early twentieth century paleontologists. *Chondrites* is a common trace fossil—usually they are not so clear as this specimen. (Value range F.)

Chondrites sp.: an especially distinctive and clear specimen of this common burrow, a trace fossil that occurs in prolific numbers in Middle Ordovician limestones. Plattin Limestone, Jefferson County, Missouri. (Value range G.)

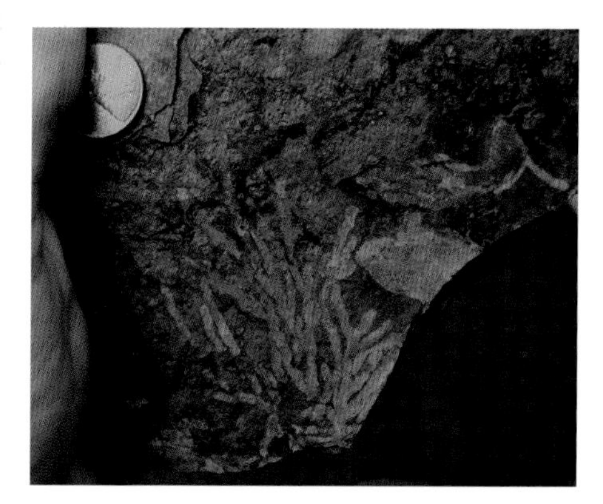

Chondrites sp.: a burrow in black shale of Pennsylvanian age that occurs above a coal seam in Illinois. It's easy to see why these would have been considered as marine plants by nineteenth century paleontological pioneers, especially when they are associated with coal seams. (Value range F.)

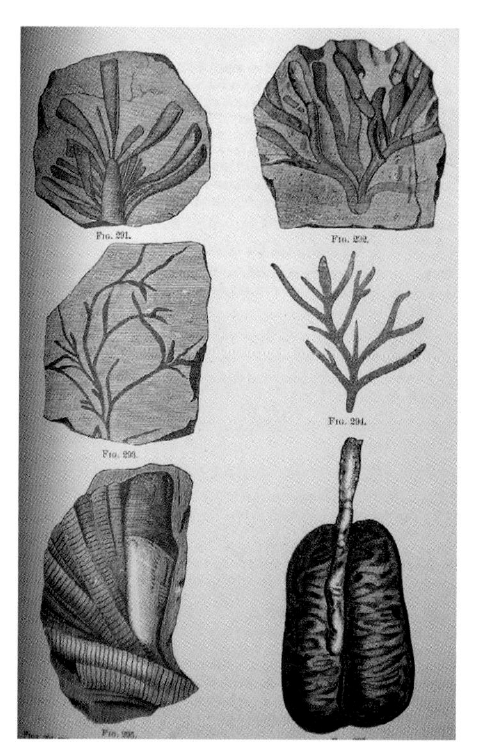

An illustration of what were believed to be fossil marine plants (seaweeds) from Joseph LeConte's 1877 *Elements of Geology.*

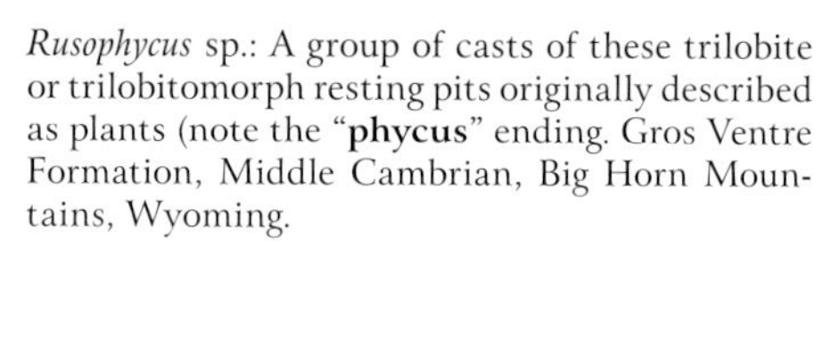

Rusophycus sp.: A group of casts of these trilobite or trilobitomorph resting pits originally described as plants (note the "**phycus**" ending. Gros Ventre Formation, Middle Cambrian, Big Horn Mountains, Wyoming.

Rusophycus sp. (*Fucoides biloba* Vanuxem, 1842): this is a sandstone cast of a burrow made either by a trilobite or a trilobitomorph (soft bodied trilobite-like animal). *Rusophycus* originally was considered as a type of sea weed (fucoid). Upper Cambrian, Lake Delton, Wisconsin. *Courtesy of Gerald Gunderson*

Conostichus sp. Lesquereux: a group of these possible jellyfish (medusa) or anemone-like animals, originally described by Leo Lesquereux as a type of sponge. Atoka Formation, Lower Pennsylvanian, Jasper, Arkansas. (Value range G, single specimen.)

Conostichus sp.: single specimen of what may be an anemone-like marine animal. Atoka Formation, Jasper, Arkansas.

Conostichus sp.: a variant on the previous, still-puzzling fossils.

Conostichus sp.: side view of previously shown specimen.

Conostichus sp.: another variant on the genus. It remains unclear if *Conostichus* and other "medusoid-like" fossils really are medusa jellyfish or anemone-like animal impressions or whether they are some type of trace fossil. What **is known** is that **they are not fossil plants, as many earlier workers considered them to be**. Hannibal Siltstone, Lower Mississippian, Hannibal, Missouri. (Value range G.)

Faux Calamites: puzzling trace fossil, which resembles a Paleozoic plant. Resembling a **calamites** (Chapter Six), this is a good example of how trace fossils can mimic a fossil plant. When it was found by the author, he thought that he had found a Devonian calamites. Sylamore Sandstone, Upper Devonian, Bentonville, Arkansas.

A Selection of Some Books on Fossil Plants

Books on fossil plants (compared to fossils in general) are limited in number. Here is a representative selection. Technical articles in scientific journals concerned with paleobotany are more common than are books, but again, compared with literature on fossil animals and protists (especially foraminifera), that on fossil plants is much smaller.

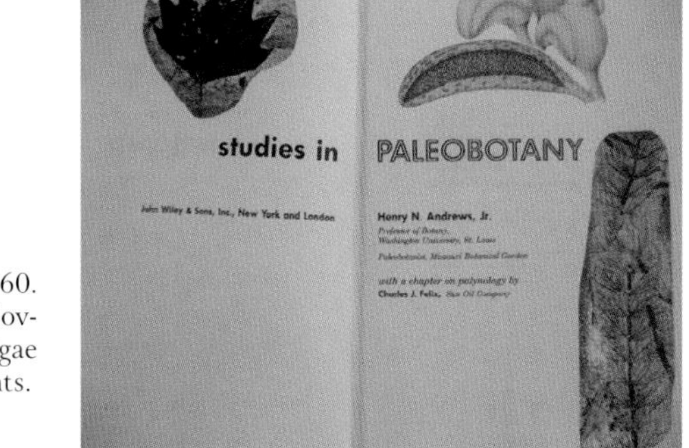

Studies in Paleobotany, Henry N. Andrews, 1960. One of the few text books on paleobotany. Covers major plant divisions and groups from algae to angiosperms. Has a focus on Paleozoic plants.

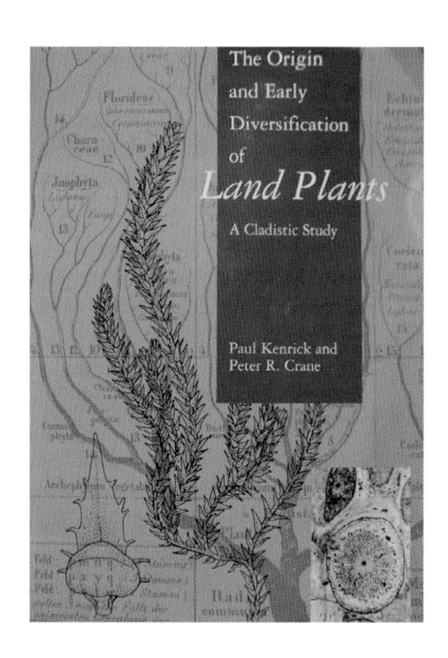

The Origin and Diversification of Land Plants: a Cladistic Study, Paul Kenrick and Peter R. Crane. Smithsonian Institution Press, 1997. A technical approach to the early fossil plant record using modern statistical techniques applied to taxonomy, which includes cladistics.

Ancient Plants and the World They Lived In. Henry N. Andrews, 1947. Comstock Publishing Co. A readable introductory text-like book on plant fossils and paleobotany. Less technical than the previous, later text by Henry Andrews.

Colorful illustrations with good coverage of a wide variety of Paleozoic fossils, including excellent material on Paleozoic plants. Specimens illustrated are predominantly from Europe.

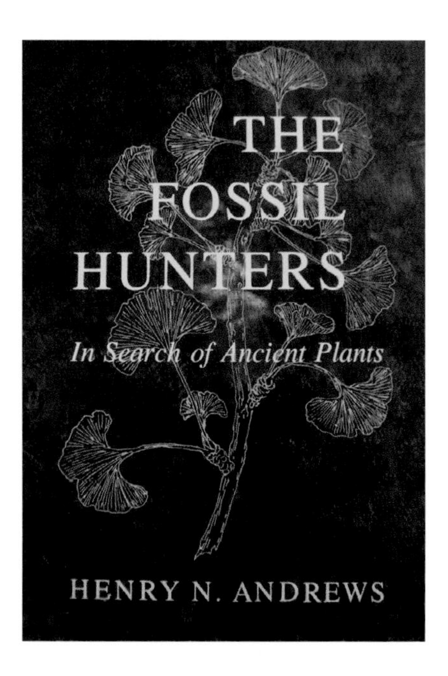

The Fossil Hunters, In Search of Ancient Plants. Henry N. Andrews. A readable yet scholarly work relating to the history of Paleobotany with an emphasis on those who not only made major discoveries in paleobotany, but who also did the necessary fieldwork to make such discoveries. (Andrews was one of the teachers in the author's education in geology. He highly recommends the work.)

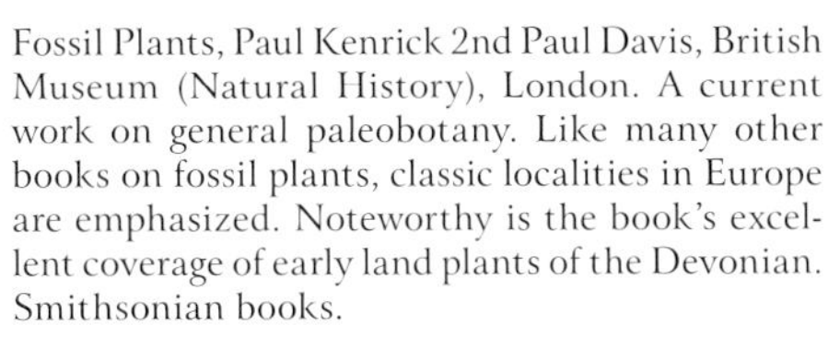
Fossil Plants, Paul Kenrick 2nd Paul Davis, British Museum (Natural History), London. A current work on general paleobotany. Like many other books on fossil plants, classic localities in Europe are emphasized. Noteworthy is the book's excellent coverage of early land plants of the Devonian. Smithsonian books.

Morphology of Plants, Harold Bold. A standard textbook on botany, with emphasis on plant taxonomy, especially those that are not flowering plants (angiosperms), the subject of most botanical texts. Fossil plants are also well covered, something that some botanists appear to ignore.

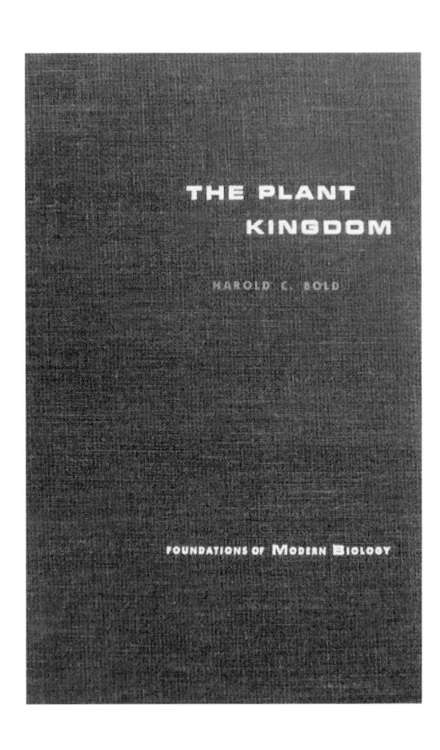

The Plant Kingdom. Harold C. Bold. A condensed version of the previously mentioned work, with fewer plant fossils.

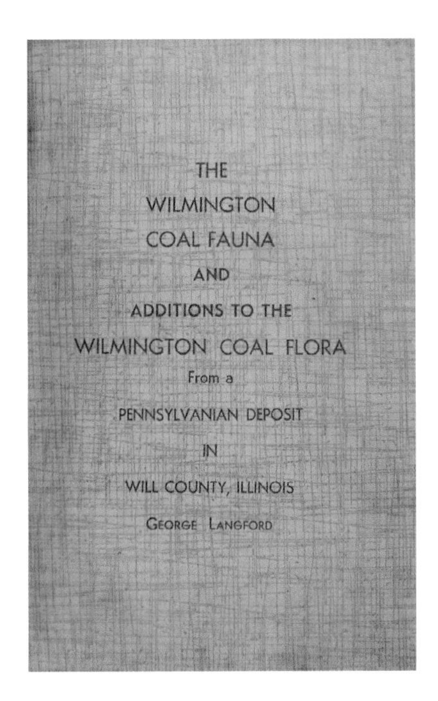

The Wilmington Coal Fauna and Additions to the Wilmington Coal Flora from a Pennsylvanian Deposit in Will County, Illinois. George Langford. Esconi Associates, 1963. A generously illustrated work on Mazon Creek fossils, which includes both plants and animals (many quite rare). Published through the auspices of the Chicago based Earth Science Club of Northern Illinois (ESCONI).

Leaves and Stems of Fossil Forests. R. E. Janssen. Illinois State Museum. An excellent, widely distributed work with an emphasis on fossil plants from the Mazon Creek fossil beds.

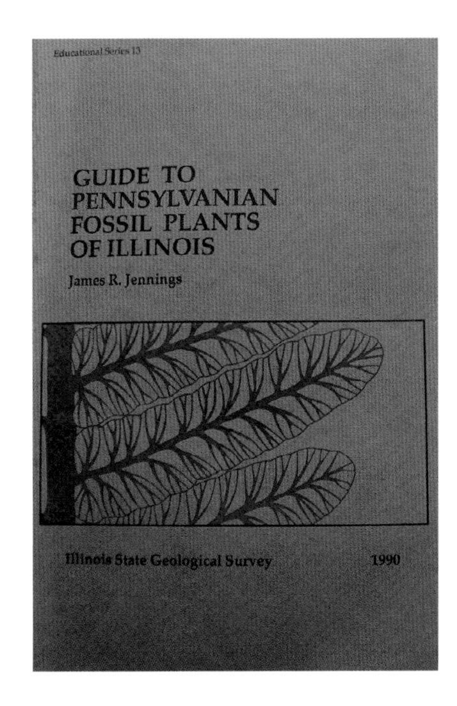

Guide to Pennsylvanian Fossil Plants of Illinois. James R. Jennings, 1990. Illinois State Geological Survey. An informative and widely distributed guide to Pennsylvanian plant fossils found in Illinois.

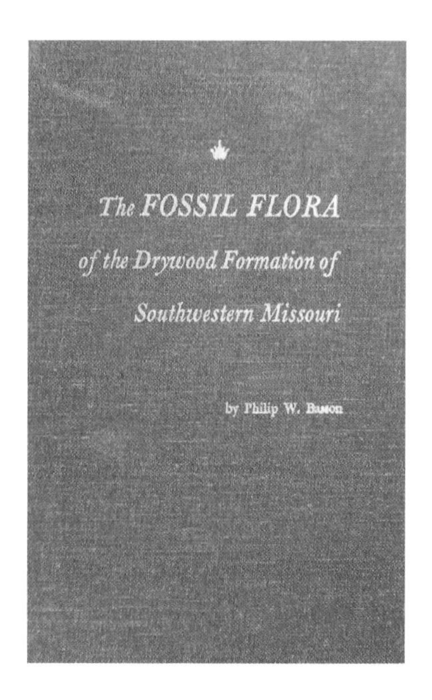

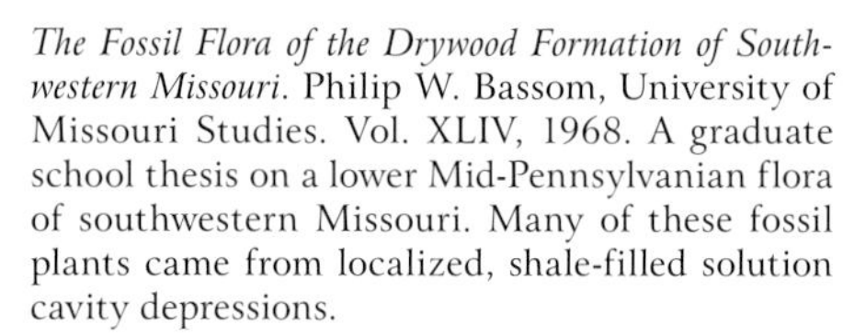
The Fossil Flora of the Drywood Formation of Southwestern Missouri. Philip W. Bassom, University of Missouri Studies. Vol. XLIV, 1968. A graduate school thesis on a lower Mid-Pennsylvanian flora of southwestern Missouri. Many of these fossil plants came from localized, shale-filled solution cavity depressions.

The Mazon Creek Fossil Flora. Jack Wittry, 2006. An attractive addition on the Braidwood facies of the Mazon Creek Lagerstatten. Published by ESCONI, Earth Science Club of Northern Illinois.

A Pictorial Guide to Fossils. Gerard R. Case, 1982. A widely distributed work that covers a broad range of fossils, but which has more on animals than on plants. Van Nostrand-Reinhold Co.

Common Fossil Plants of Western North America. 2nd. Edition. William D. Tidwell. A comprehensive work on fossil plants found in the western states of the U.S. As Cenozoic and Mesozoic strata are especially well represented in the "mountain states," there is something of a bias toward angiosperms and conifers, less on Paleozoic plants.

Common Fossil Plants of Western North America. The first edition of an interesting work with a focus on collecting plant fossils.

Secrets of Petrified Plants. Ulrich Dernbach and William D. Tidwell, 2002. A coffee table book that deals with attractive plant petrifications, especially colorful Permian petrified tree ferns of France, Brazil, and elsewhere. It deals also with coal balls of the U.S. Midwest. The book is colorful, attractive, and interesting. D'Oro Publishers, Germany.

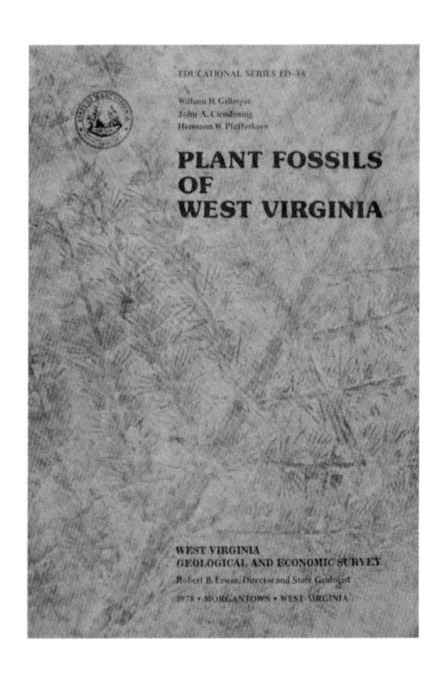

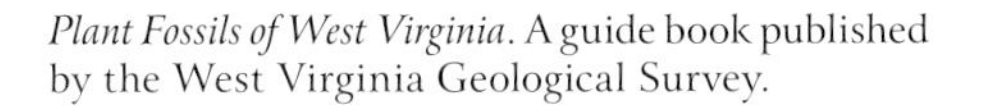
Plant Fossils of West Virginia. A guide book published by the West Virginia Geological Survey.

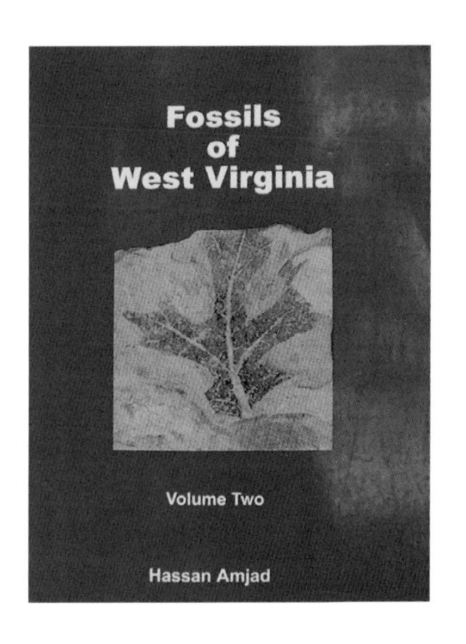

Fossils of West Virginia, Vol. 2. Hassan Amjad. A look at Pennsylvanian fossil plants from a state almost entirely underlain by non-marine rocks of this age.

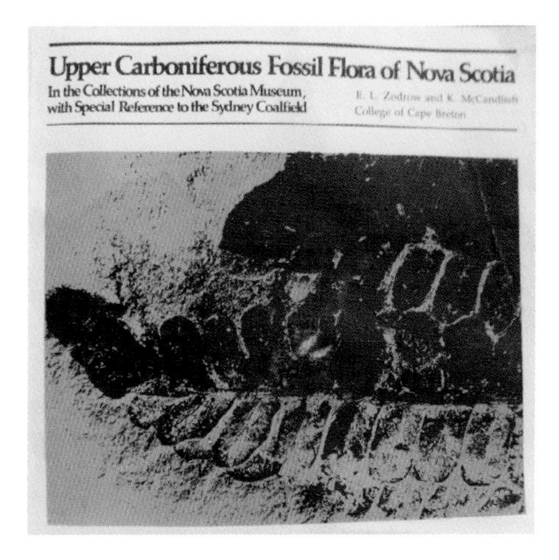

Upper Carboniferous Fossil Flora of Nova Scotia: In the Collections of the Nova Scotia Museum with Special Reference to the Sidney Coalfields. E. L. Zodrow and K. McCandlish, College of Cape Breton, 1980. Published by the Nova Scotia Museum, Halifax, Nova Scotia. Covers Upper Carboniferous fossil floras that are almost identical to those of Wales and England.

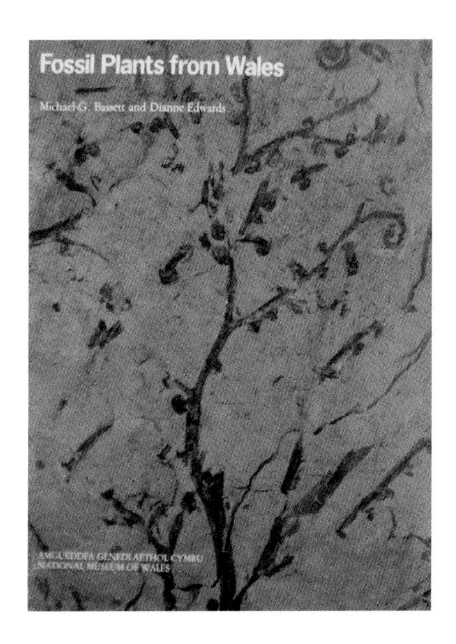

Fossil Plants from Wales. Michael G. Bassett and Dianne Edwards. National Museum of Wales. Includes some interesting material on Devonian and Lower Carboniferous fossil plants.

Geology of Alabama. Geological Survey of Alabama. This widely distributed book has a section on late Paleozoic fossil plants of Alabama.

Biostratigraphy of Fossil Plants. A technical work that uses fossil plants as a means to index strata in a manner similar to the better known use of fossil animals.

Nineteenth Century Works on Fossil Plants

Nineteenth century works on paleontology and paleobotany are sometimes available and although these classic works can often be quite technical, they can also be interesting for a number of reasons, including the fact that they represent pioneering works in geology and paleontology. Shown here are some dealing with Paleozoic fossil plants.

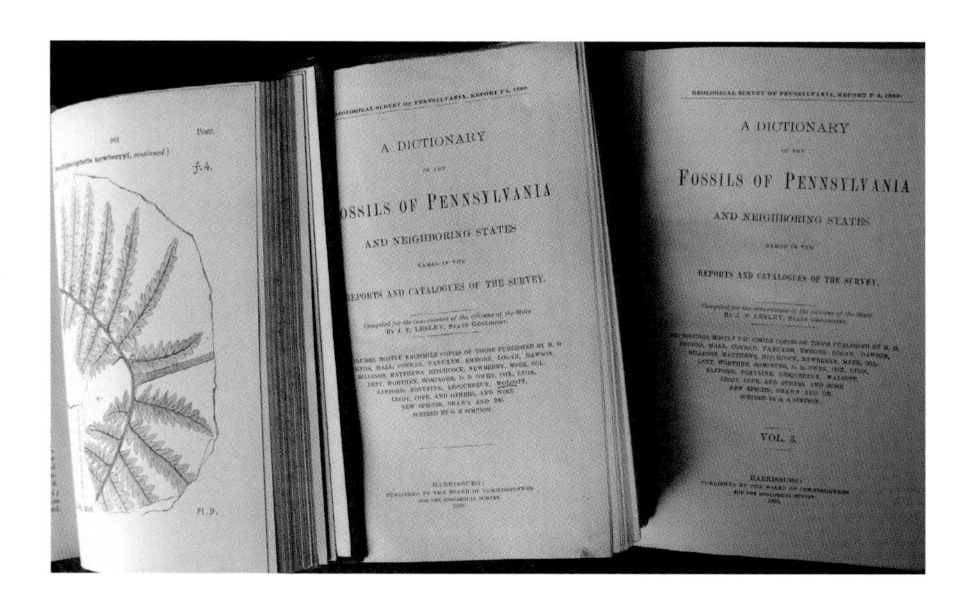

A dictionary of *The Fossils of Pennsylvania and Neighboring States* named in the reports and catalogues of the survey. J. P. Lesley, 1890. A comprehensive and curious work that attempted to both illustrate and describe **all fossils described in the scientific literature of the eastern states bordering Pennsylvania**. The work is a gold mine for someone wanting to probe more deeply into nineteenth century American paleontology and paleobotany. Many Paleozoic plants (and what were thought to be plants) are covered in this three volume work.

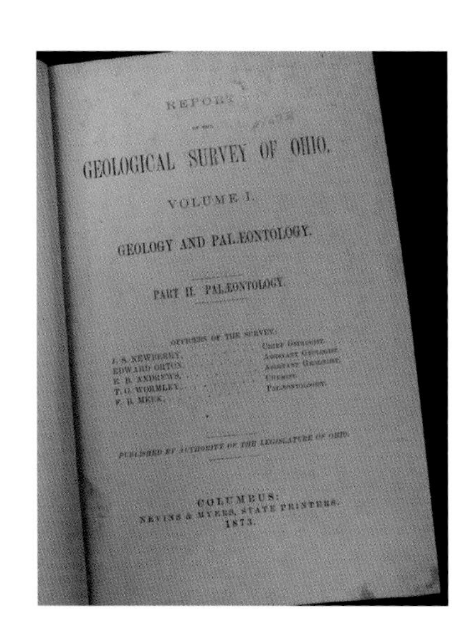

Report of the Geological Survey of Ohio. Vol. 1. Geology and Paleontology. J. S. Newberry. One of a number of classic comprehensive nineteenth century works on the paleontology of Ohio. Includes a considerable amount of information relating to Paleozoic fossil plants with which Newberry was quite familiar.

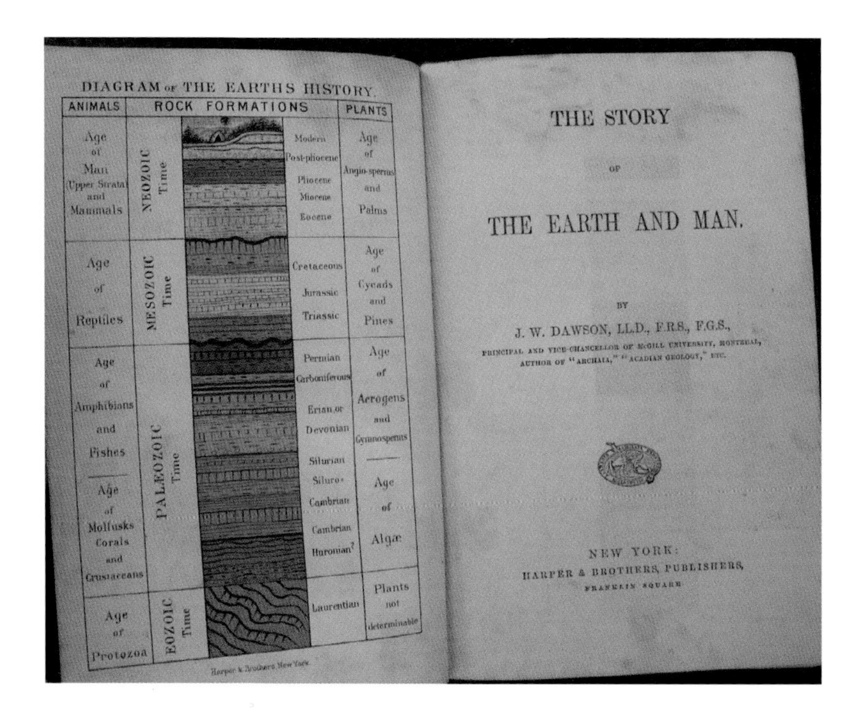

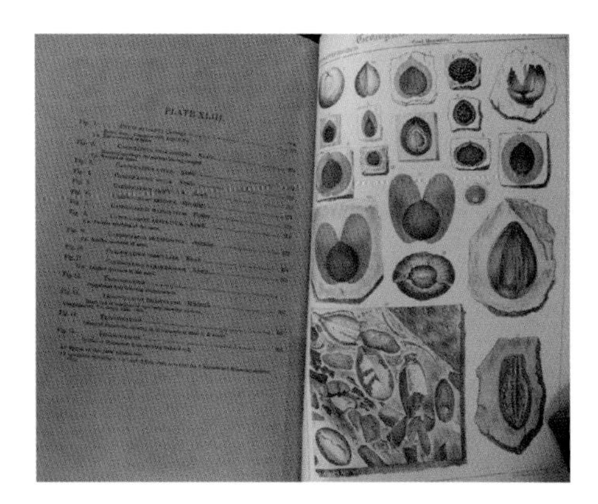

Representative paleobotanical plate from the previously mentioned work.

Story of the Earth and Man. James William Dawson, 1873, McGill University. Covers, in an interesting manner, a considerable amount of material dealing with fossil plants known in the nineteenth century. Dawson did extensive early work on fossil plant floras, especially those of Devonian and Carboniferous age in the Canadian Maritimes. The work is also interesting because Dawson was a creationist who did not accept Charles Darwin's Evolutionary Theory. Dawson was also the discoverer of Eozoon, the first "fossil" to be found in Precambrian rocks, where Darwin's theory of evolution required there to be fossils of life forms that predated the Cambrian Period. Many creationists in the nineteenth century considered that the sudden appearance of Cambrian flora and fauna found in the fossil record represented God's creation of life.

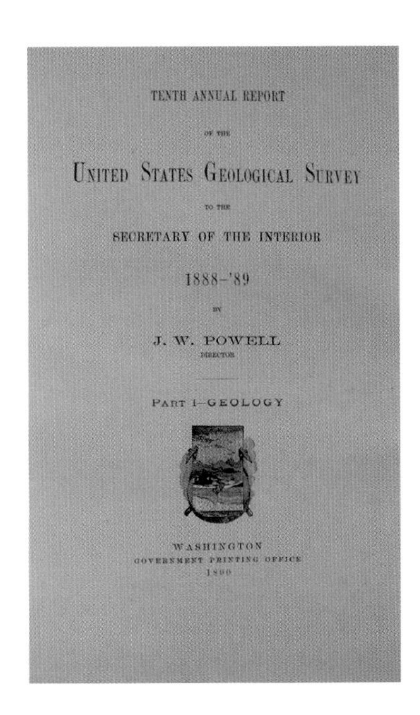

TENTH ANNUAL REPORT

UNITED STATES GEOLOGICAL SURVEY

SECRETARY OF THE INTERIOR

1888–'89

J. W. POWELL

PART I—GEOLOGY

WASHINGTON
GOVERNMENT PRINTING OFFICE

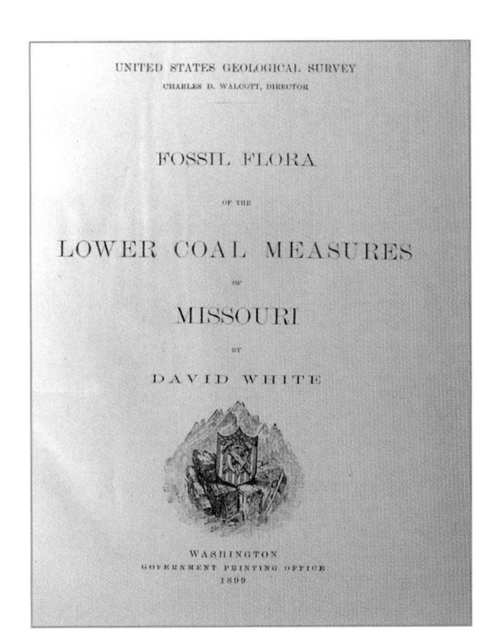

UNITED STATES GEOLOGICAL SURVEY
CHARLES D. WALCOTT, DIRECTOR

FOSSIL FLORA

LOWER COAL MEASURES

MISSOURI

DAVID WHITE

WASHINGTON
GOVERNMENT PRINTING OFFICE
1899

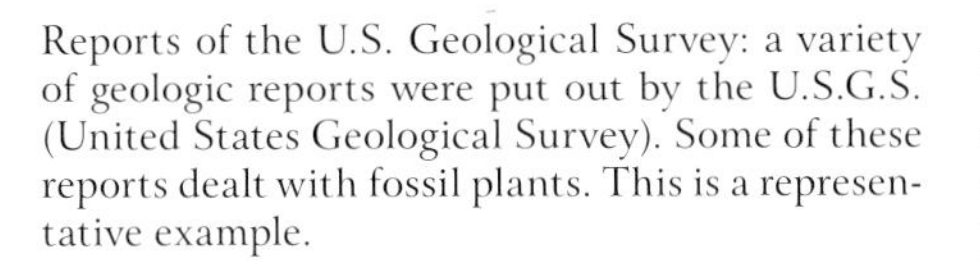

Reports of the U.S. Geological Survey: a variety of geologic reports were put out by the U.S.G.S. (United States Geological Survey). Some of these reports dealt with fossil plants. This is a representative example.

Fossil Flora of the Lower Coal Measures of Missouri. David White, U.S.G.S. Monograph. One of a number of late nineteenth century monographs dealing with fossils, including fossil plants. Those plant fossils illustrated and described here came from an exceptional flora found associated with local coal beds in western Missouri. Many of these plant fossils were collected by John H. Britts of Clinton, Missouri, and mentioned by the author.

REPORT

GEOLOGICAL SURVEY

WISCONSIN, IOWA, AND MINNESOTA;

A PORTION OF NEBRASKA TERRITORY.

DAVID DALE OWEN,

PHILADELPHIA:
LIPPINCOTT, GRAMBO & CO.
1852

Report of a Geological Survey of Wisconsin, Iowa, and Minnesota. David D. Owen. An extensive and classical work published in 1852 on the geology (and paleontology) of Wisconsin, Iowa, and Minnesota. (It also covers some of the geology of northern Missouri and Nebraska.) The report is based upon extensive geological fieldwork (much by canoe) done in the 1830s and '40s to determine local geology of these territories prior to statehood. It appears to have been a widely distributed work and still can be seen with some frequency considering the pre-Civil War date of its publication. It also is still useful as a reference to the geology of the areas covered. The author has used it successfully to find some fossil localities mentioned in the work exposed along rivers in Iowa, Missouri, and Minnesota.

Bibliography

Hantzschel, Walter, 1975. *Miscellanea, Trace Fossils and Problematica. Treatise on Invertebrate Paleontology.* Geological Society of America and the University of Kansas, Lawrence.

Glossary

Medusoids: a jellyfish or jellyfish-like animal of the Phylum Cnidaria. Radial impressions of various types often are found on bedding surfaces of both marine and non-marine strata. Many of these have been considered as impressions of jellyfish (or medusa-like) animals. Such animals in modern oceans can wash up onto a beach and, upon desiccation, leave a circular impression in the beach mud or sand. The origin of these medusoid-like impressions can be from many sources besides actual jellyfish impressions. These can include such things as small sand "boils," concretion impressions, sea weed impressions, and burrow entrances.

Trace Fossil: a track or trail made by some—usually soft-bodied—type of animal. Many trace fossils originally were described as fossil seaweeds (fucoids), which they do resemble in many cases.

Treatise on Invertebrate Paleontology: an exhaustive work jointly published by the Geological Society of America and the University of Kansas. This series of books attempts to illustrate and describe all genera of invertebrate animal fossils, including trace fossils.

Chapter Nine
Forests in Transition

by Tim Northcutt

It's a bit as though I entered the world of paleontology through the back door, late for an important lecture, and have been trying to find out what I missed ever since. But that has been my challenge. I actually discovered paleontology during graduate school in the visual arts through chance introductions with two people from different avenues of my life. Both of them had interest in paleontology and I was quickly introduced to fossil collecting. For a few months we collected locally, then soon in the Cretaceous Smoky Hill Chalk Beds of western Kansas. As my basic collecting skills advanced and curiosity level grew, sites that were further from home, but still within a few hours drive, became immediate targets, such as the Tonganoxie plant fossils.

Tonganoxie pteridospermophytes or pteridosperms

These are the seed ferns. They were extinct plants that arose on upright trunks bearing helically arranged massive fern-like fronds. Some seed ferns have large solitary seeds. Others have small multi-ovulate capsules.

Nearby targets also included the beautifully preserved Upper Pennsylvanian flora from Franklin County, Kansas. The finer grained sandstones found here—just above the coal that separates the Tonganoxie Sandstone from the Weston Shale below—preserved a diverse flora of Upper Pennsylvanian taxa. A very high degree of preserved relief often distinguished the flora, providing a natural beauty that drew many collectors, and even authors (Andrews 1947, see previous chapter), to the specimens. Unfortunately there are currently no practical locations to collect this much sought after flora. The Tonganoxie Flora includes a substantial group of ferns (pteridophytes), seed ferns (pteridosperms), clubmosses (lycopods), horsetails (Sphenophytes), and cordaites. It was the first of two fine Upper Pennsylvanian floras that I've had the good fortune to collect.

Alethopteris grandinii: from the Tonganoxie Sandstone near Ottawa, Kansas. This seed fern perhaps best exemplifies the conspicuously three dimensional preservation of the Tonganoxie Sandstone Flora.

Sphenopteris sp.: from the Stranger Formation, Tonganoxie Sandstone, Ottawa, Kansas.

Neuropteris scheuchzeri Hoffman: from the Stranger Formation, Tonganoxie Sandstone near Ottawa, Kansas. Note the two leaflet morphs.

A rare element of the Tonganoxie Sandstone Flora, *Odontopteris* sp. When I found this specimen, it would be years until I rediscovered this genus at Parkville, where it is the most common element in the flora.

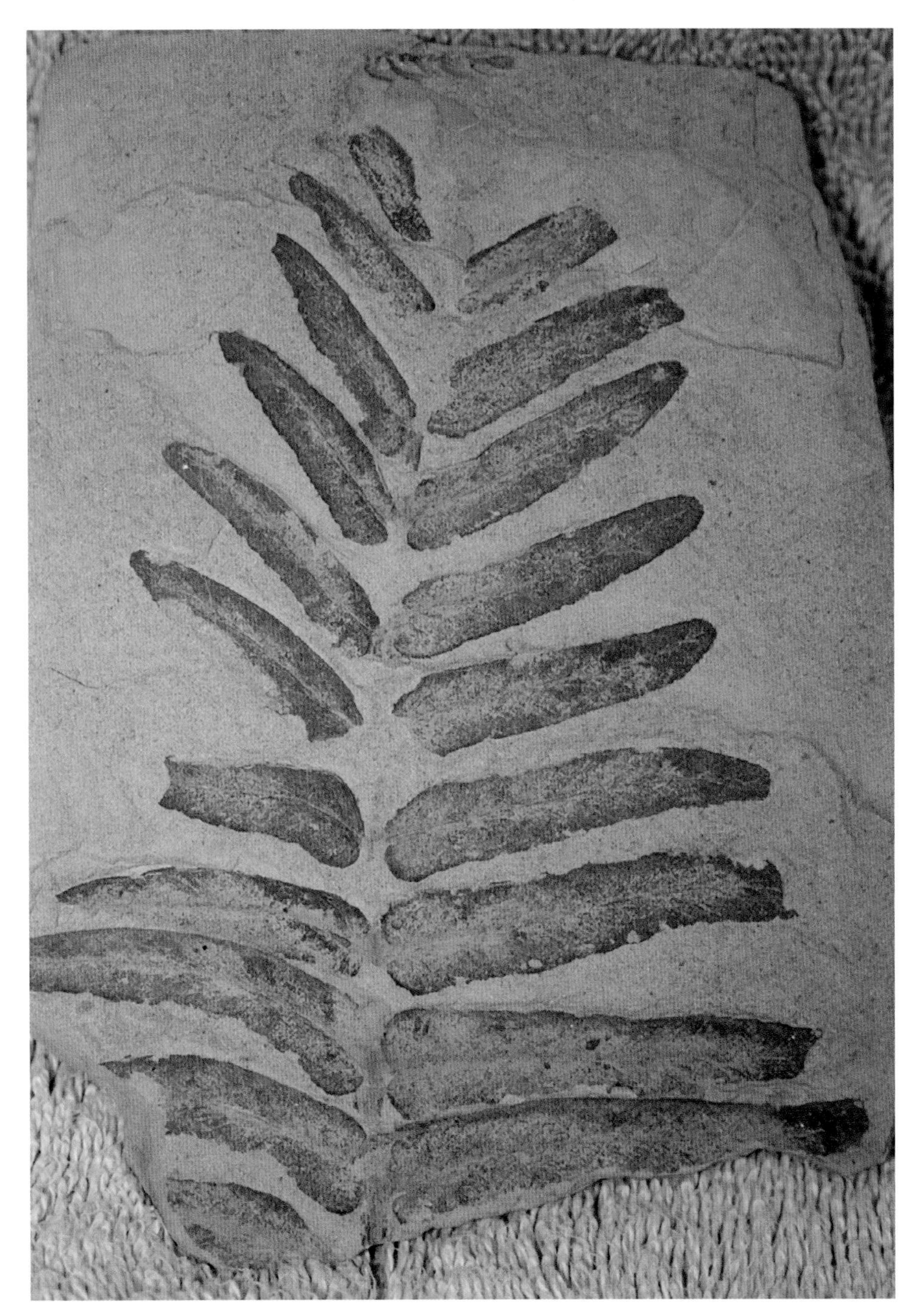

Neuropteris scheuchzeri from the Stranger Formation, Tonganoxie Sandstone of Ottawa, Kansas.

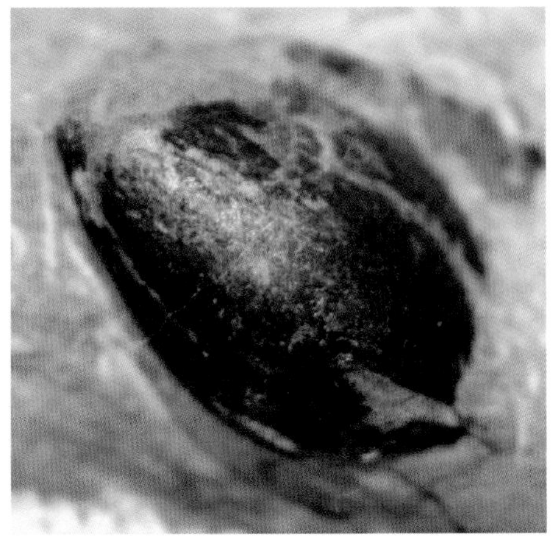

Trigonocarpus sp. from the Stranger Formation, Tonganoxie Sandstone of Ottawa, Kansas. *Trigonocarpus* is typically the largest solitary seed found in Pennsylvanian floras.

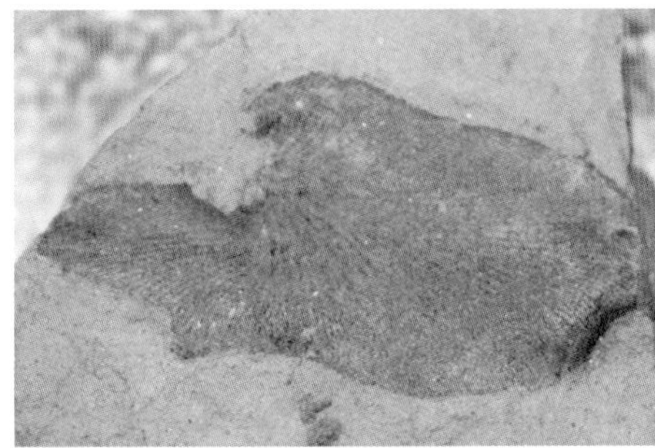

Cyclopteris sp. from the Stranger Formation of Ottawa, Kansas. Note the smooth, or entire, margin.

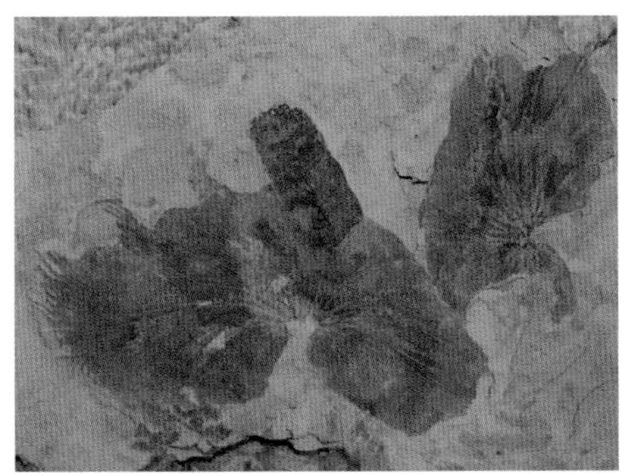

Two cyclopterid-like aphlebia from the Stranger Formation, Tonganoxie Sandstone, Ottawa, Kansas. Note the incised margin at the left indicating the margin is missing, or incomplete elsewhere. Note also these are not solitary, but rather serially arranged leaflets.

Parkville Pteridospermophytes

The other example, the Parkville Flora, was discovered during excavation at an area college. The Parkville site soon became a fairly well known locally. Like many fossil sites, it requires serious effort and devotion to extract the highest quality specimens. These specimens are the ones buried deep below the surface and undamaged from the freezing and thawing of the winter months, and then the drying and shrinking of summer. The Parkville Flora is perhaps best represented by the seed fern *Odontopteris brardii* (Brongniart). Though found throughout the strata, a particular lens has produced a wealth of large specimens. Michael Krings (et al 2006) offered an update to the collective knowledge of the leaf architecture of *Odontopteris brardii*, largely based upon specimens collected from that lens. Handsome specimens of highly dissected pinnules, or leaflets, presumed to be associated with Odontopteris, are also present. Of particular note in the flora is *Walchia* sp., generally considered as one of the first conifers. The Parkville flora also has produced a spectacular array of fruiting bodies, which will likely bring further attention from the academic community. Among them is a collection of specimens discovered that is believed to be a pollen organ with affinities to the lyginopterid *Feraxotheca*, a genus previously unknown from the Upper Pennsylvanian.

For those less familiar with Kansas paleontological history, both the state and the paleontological community have benefited greatly due to the wealth of fine late Paleozoic fossils. For decades, regional museums have proudly displayed specimens from the Pennsylvanian lagerstatten at Hamilton and Garnett, offering paleontological enlightenment via their distinctive and defining flora and fauna. Items such as the oldest known diapsid reptile, *Petrolacosaurus*, giant millipedes, rare Pennsylvanian conifers (*Walchia*), sea scorpions, and early sharks have all lent insight into the distinct biota of Pennsylvanian strata of the mid-continent. Why is the Parkville flora so special? The fine grained mudstones and sandstones laid down in quiet backwaters of a floodplain environment preserved many of the finest details of the plant's morphology, as well as insects and amphibian trackways, leading some to ponder whether it should be included among the lagerstatten of Kansas. Form species representing nearly every avenue of Upper Pennsylvanian plant life from seed ferns (*Neuropteris ovata, Sphenopteris* sp., and *Aphlebia* sp.), lycopod trunks, branches, leaves, & cones, fertile & sterile fern fronds (*Pecopteris candolleana, P. miltoni* & *Asterotheca* sp.) of true ferns, cordaites and (*Cordaianthus* sp.) incertae sedis, and a wide variety of horsetails (*Palaeostachys, Asterophyllites equisetiformis* & *longifolius, Annularia sphenophyllites, Annularia* sp., *Calamites cisti, Calamites* sp. & *Sphenophyllum oblongifolium*) make it a special place for both amateur and professional paleobotanists.

A handsome specimen of the seed fern *Sphenopteris* sp. from the Bonner Springs Shale, Parkville Flora.

Neuropteris ovata from the Bonner Springs Shale at Parkville, Missouri. Leaflets approx. 3.5 cm.

Odontopteris brardii from the Bonner Springs Shale at Parkville, Missouri. The poster boy for the Parkville Flora is an uncommon element at best in most Upper Pennsylvanian floras, making the Parkville flora special, if just for this circumstance alone.

The terminal end of a frond of highly incised pinnules, or leaflets, of *Aphlebia* sp., presumed to be associated with *Odontopteris brardii,* noteworthy for its terminal aspect.

Other Pteridospermophytes

Familiarity with the Tonganoxie and Parkville floras led me and my friends to search out other regional plant sites so that we might better understand where these floras stand within the greater scope of Pennsylvanian plant life. One such site that garnered our interest is a site near Knob Noster, Missouri (Chapter 4). Though it has been in private hands since its discovery and mined by the landowners, their marketing of the collection has made the biota well known and sought after. The Knob Noster flora is preserved (in part) in a fashion similar to that of the famous Mazon Creek biota of Illinois, with both plants and animals preserved in siderite concretions.

Not far from Knob Noster, at Warrensburg, is another Middle Pennsylvanian site that produces ferns and sphenophytes from the Pleasanton Group indicative of the floras examined by David White from the historic coal beds of west-central Missouri during the early twentieth century.

No two fossil sites ever seem to exhibit either the same flora or preservational characteristics. The Middle Pennsylvanian Cabaniss Formation exposure at Oswego, Kansas, certainly presents an example of this. Extensive excavation rarely produces anything but disarticulated plant parts (*Neuropteris scheuchzeri* Hoffman & *Annularia stellata* Schlotheim). Yet the surviving parts are well-preserved—until metallic elements (marcasite) in the strata oxidize and destroy everything nearby. That might be a bit aggravating were it not for the lure of finding one of the tiny starfish also found in this strata. Further afoot from home is a Lower Pennsylvanian site in Brown County, Illinois, with elements much different from the more familiar later Pennsylvanian flora near home. A single day in the snow one March afternoon failed to produce any sign of the coveted *Lacoea*, a bizarre lycopod fruiting body resembling a small sunflower head after the petals have fallen. Still, even with snow on the ground and more falling, a few small examples peeked out from below the covering of white to suggest what might be there on a pleasant day (*Alethopteris* sp. & *Samaropsis* sp.). Pennsylvanian coal beneath the surface in eastern Illinois caught fire and burned, producing the "clinker" specimens of lycopod trunks that caught my eye years ago at MAPS Expo in Macomb, Illinois. Additionally, the clinker (or burnout) offerings included a pecopteris fern and a lycopod root cast. If you've been to even a few fossil exhibits, you've likely seen the plates of hard black shale from St. Clair, Pennsylvania, covered with *Alethopteris* fronds and a few other taxa. Nearby Pottsville mines produced the wide variety of lycopod trunk specimens. Fossil specimens of tree trunk, or bark, present challenges to those wishing to identify them, for these trees often possessed several layers of tissue. Each of these layers may potentially be preserved during different stages of the plant's decomposition.

Trigonocarpus sp.: a large seed fern fruit from the Pennsylvanian at Knob Noster, Missouri. Seed approximately 3 cm.

Neuropteris sp.: a large fern pinnule from the Mazon Creek concretionary fauna southwest of Chicago, Illinois. Leaflet approximately 6 cm.

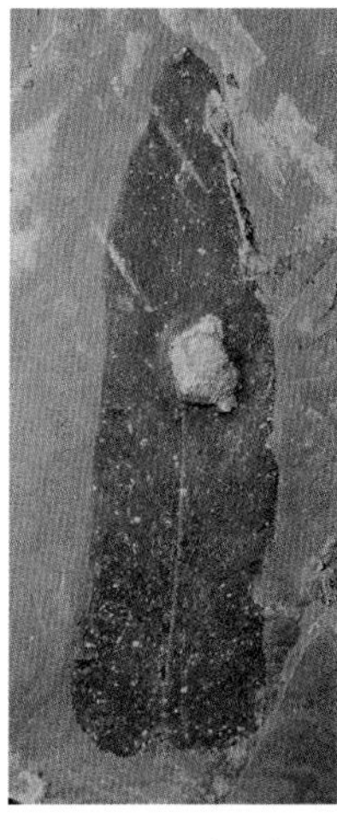

Neuropteris scheuchzeri Hoffman: Alarge pinnule or leaflet of a seed fern from the Cabaniss Formation at Oswego, Kansas.

Alethepteris sp.: a single pinnule from a seed fern found in Brown County, Illinois, one March when snow still covered the ground.

Winterset Mystery Ferns

Often overlooked and under appreciated locally have been some ferns found in the Winterset Formation at several localities around Kansas City. Reasons behind the neglect seem to stem (in part) from the difficulty most have had in identifying the ferns. Everywhere the Winterset Flora presents itself, it includes a very similar looking mix of plant life. It has been years since I recognized that all the ferns in the Winterset Formation belong to the neuropterids, generally a fern with relatively large pinnules, or leaflets; however, one form has very small pinnules, the other more typical, larger leaflets. When Bruce called with the intent of enlisting me in this project, he mentioned one group of fossil plants from Kansas City that had puzzled him, the Winterset flora. Immediately, I acknowledged a similar puzzlement and accepted the challenge of clearing up this long neglected flora. As I perused the typical references looking for clues to their identification, I realized one particular neuropterid that was consistently illustrated at several times the actual size, allowing the pinnules to appear similar in size to many of the other neuropterids shown. "*Neuropteris heterophylla*" the identifications read. Realizing the etymological inference, I immediately wondered if I could be onto something. It had occurred to me that the ferns included in the Winterset could all belong to the same plant. The small pinnuled variety, in my mental model, formed the body of the plant and the cardio-formed pinnules were aphlebia-like morphs or basal pinna. When I carried my search further, I began to doubt my original conclusions. Satisfied that I had exhausted my limited immediate resources, I contacted Rudy Serbet, curator of paleobotanical collections at the University of Kansas. Rudy has been a good liaison between the amateur community and academia since his arrival at the university. He suggested we send some images to Michael Krings, the German paleobotanist who led a project that documented aspects of *Odontopteris brardii* based largely on specimens I collected at Parkville in the late eighties. Michael responded in regard to the larger pinnuled ferns immediately. Backing his conclusion that the larger pinnules belong to the form species known commonly as *Neuropteris scheuchzeri*, he sent a reference. Foliar forms of *Macroneuropteris scheuchzeri* (Pennsylvanian, Sydney Coalfield, Nova Scotia, Canada), by Erwin L. Zodrow, documents a wide variety of foliar forms he attributes to *Macroneuropteris scheuchzeri*, an arborescent-medullosan fern species erected in 1999 (Cleal & Shute erected genera in 1995). I was a bit taken aback. Over the years I'd come across many examples of *Neuropteris scheuchzeri*. They seemed to be found at most places where Pennsylvanian plants are found and all of them looked alike, or so I thought. All the specimens I knew were convex at the apical or distal end. The large pinnuled fern from the Winterset was cardio-shaped, thin and concave at the apical end. And though the attribution of the smaller pinnuled morph remains less than certain, for now I'll trust my research, which indicates the smaller pinnuled fern fits best within the possibly dubious *Neuropteris heterophylla*—satisfactory until I learn better.

Neuropteris "Macroneuropteris" heterophylla: an example of a uniformly small pinnuled neuropterid specimen from the Winterset Limestone, near Kansas City, Missouri. These seed ferns have often been mistaken for pecopterids, but do not have the constricted attachment; rather, it is broad with proximal ends of leaflets overlapping the rachis.

A specimen exhibiting a mix of pinnule forms from the Winterset Limestone near Kansas City, Missouri.

As with most such inquiries, the unexpected discoveries have proven the most valuable. Amateurs who are determined do indeed have opportunities to make worthwhile contributions to science. Now, just over twenty years after first discovering the rich paleobotanical resources available nearby, my research has led to numerous actions being taken by professional paleontologists to address the issues I've presented. Unresolved issues are everywhere in paleontology. One only needs to identify the unresolved issue and then set about collecting information both in the form of data and physical specimens, to extend the envelope of science.

The element of camaraderie within the paleontological community is not one to be dismissed and the example from the Winterset Limestone is a reminder of that. A friend whose life has been one challenge after another rode the bus to collect a large patch of Winterset plants that included a bunch of cordaite seeds. Despite the effort necessary to get them and get them back home, he shared this example with me years ago. Responsible citizenship is a practice to which one should always adhere. However, fossils that have reached the surface, such that any part of them is visible, means that they soon will meet destruction by Mother Nature's wrath. Those who claim to want to protect fossils by leaving them in the ground are foolishly dooming them to an unnecessarily early demise. Learn what you can from the fossils you acquire and share the knowledge with all who are interested.

Neuropteris scheuchzeri: another frond exhibiting the elongated form from the Winterset Limestone near Kansas City, Missouri.

Neuropteris scheuchzeri, a frond exhibiting the elongated form of pinnule with a pointed apex from the Winterset Limestone near Kansas City, Missouri.

Pteridophyta

These are the true ferns, which bear sporangia marginally or abaxially (on the underside of leaflets) on the fronds and the spores germinate to form viable gametophytes.

These are the true ferns. Pteridosperms were not really ferns.

Sterile frond of a true fern, *Pecopteris candolleana*. Although common in many Middle Pennsylvanian floras, pecopterids are few and far between in the Parkville Flora.

Pecopteris sp.: a fern from the Mazon Creek concretionary fauna southwest of Chicago, Illinois.

Another sterile frond of a fern, *Pecopteris miltoni*, of the Parkville Flora (Bonner Springs Shale).

Asterotheca sp.: fertile fern fronds from the Bonner Springs Shale at Parkville, Missouri. Note the paired sporangia on the pinnules.

Pecopteris sp.: a fern from the clinker (or burnout) beds of Danville, Illinois. "Clinker" or burnout refers to the overlying clay or shale naturally fired when beds of coal below burn.

Pecopteris pseudovestita White: a fern from the Middle Pennsylvanian Pleasanton Group near Warrensburg, Missouri.

Alloiopteris sp.: from the Stranger Formation of Ottawa, Kansas. Note the 5–6 cm long pinnules.

Lycophyta or Lycopodophyta

These are the scale trees. The Division name is variously given as both Lycophyta and as Lycopodophyta.

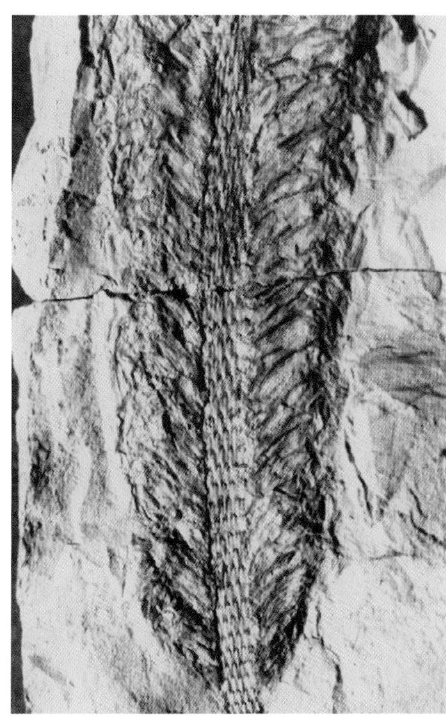

Lepidostrobus princeps Lesquereux: a lycopod fruiting structure, or cone, from the Bonner Springs Shale at Parkville, Missouri. Note the helically arranged sporophylls.

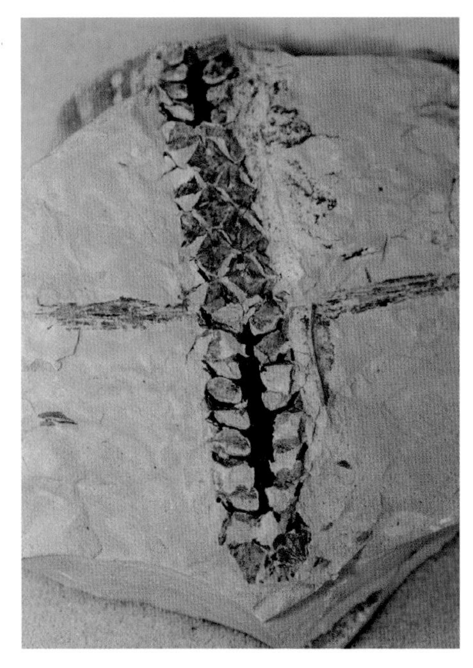

A partial specimen of *Lepidocarpon* sp., a long hanging cone of a lycopod, which may reach lengths of more than half a meter.

Lepidodendron sp.: from the Bonner Springs Shale at Parkville, Missouri. Though pieces of their narrow strap leaves with prominent midveins are found somewhat commonly in the flora, trunk specimens are rare finds in the Parkville Flora.

Lepidodendron sp. from the Bonner Springs Shale at Parkville, Missouri.

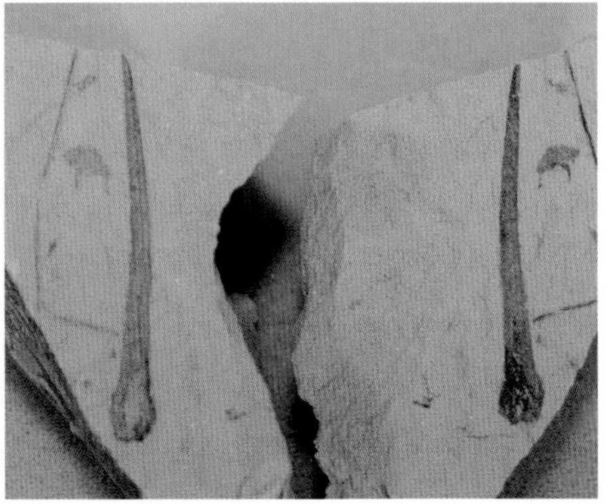

Young leaf of a lycopod, *Lepidophylloides* sp.

Lepidophloios sp.: a lycopod trunk from the clinker beds of Danville, Illinois. Clinker beds (or burnouts) form when coal beds catch fire and fire the shale above them like pottery clay in a kiln.

Knorria sp.: a cast of another lycopod trunk from the clinker beds of Danville, Illinois. *Knorria* represents a state of decortication of *Lepidodendron* in which most all of the tissues external to xylem was lost.

Stigmaria sp.: part of a lycopod root cast from the clinker or burnout beds of Danville, Illinois.

Asolanus sp.: a lycopod trunk from the home of the American Upper Carboniferous, Pottsville, Pennsylvania. *Asolanus* represents a particular decortication stage of lycopod trunks.

Knorria sp.: a portion of a lycopod trunk from Pottsville, Pennsylvania.

Sigillaria sp.: a distinctive lycopod root that displays its attachment scars in vertical columns from the Lower Pennsylvanian of Pottsville, Pennsylvania.

Lepidophloios sp.: a lycopod trunk from Pottsville, Pennsylvania.

Sphenophytes

These are the Arthrophytes of Chapter Six. Living forms are commonly known as horsetails or scouring rushes. They are distinguished by clonal growth characterized by extensive underground rhizome system from which rise upright axes bearing whorls of leaves and branches at the nodes.

Sphenophyllum oblongifolium: from the Stranger Formation of Ottawa, Kansas. Note the well preserved venation.

Calamostachys sp.: from the Stranger Formation of Ottawa, Kansas. Note the bracts that spread at a nearly right angle from the axis.

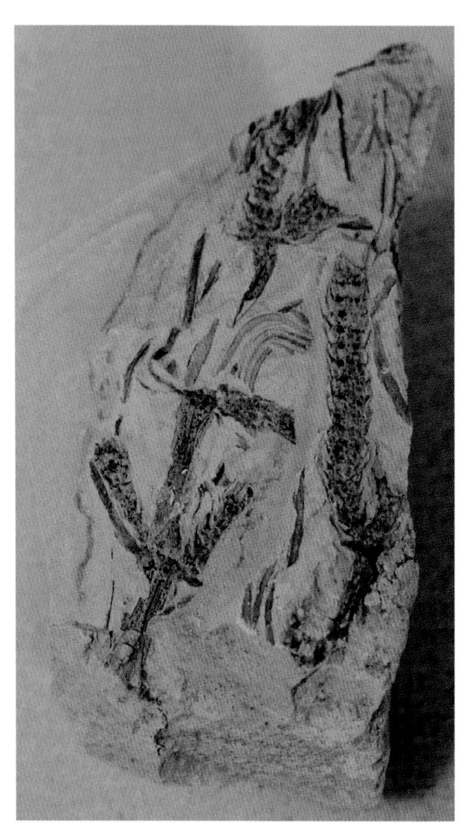

Paleostachya sp.: from the Bonner Springs Shale at Parkville, Missouri. Note the sporangia on the right cone and bracts, which bend nearly parallel to the axis.

Palaeostachya sp.: a cone of one of the Parkville Floras horsetails.

A nicely articulated horsetail, *Annularia* sp., from Parkville, Missouri. Note how the whorls of **Annularia** leaflets do not overlap.

Annularia sp.: an example of one of the horsetails from the Middle Pennsylvanian of Knob Noster, Missouri.

Asterophyllites equisetiformis: from the Bonner Springs Shale of Parkville, Missouri. (Value range F)

Asterophyllites longifolius: from the Bonner Springs Shale of Parkville, Missouri. Note the particularly long foliage.

Annularia sphenophylloides: from the Bonner Springs Shale of Parkville, Missouri. Though they look flower-like, they predate the flowers by tens of millions of years.

A colorful specimen of the vine-like, herbaceous horsetail *Sphenophyllum oblongifolium*.

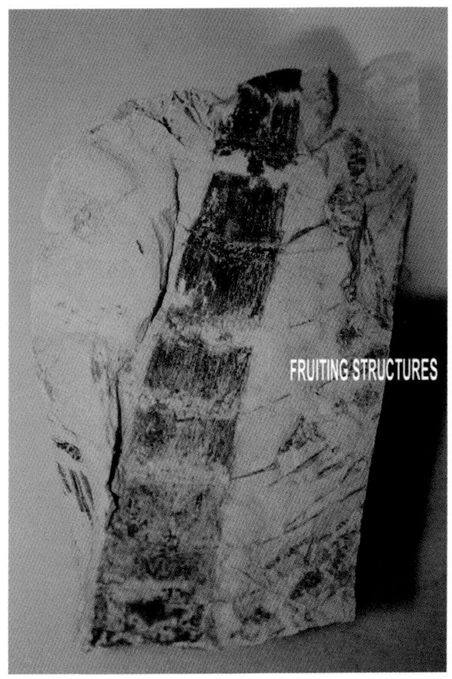

Calamites sp.: with fruiting structures arranged to the right borne out of the nodes.

Calamites cisti: from the Bonner Springs Shale of Parkville, Missouri. Note the three-dimensional preservation of the trunk.

Asterophyllites equisetiformis: an example of another Middle Pennsylvanian horsetail from Knob Noster, Missouri.

Cordaites, Coniferphytes and Incertae Sedis

Cordaites arise as single trunked arbors with a crown of large strap-like leaves. Some likely grew to more than 45 meters tall. Pollen and seed producing organs consist of compound cones, generally borne on distal leaf-bearing branches.

Coniferphytes generally are evergreen plants.

Incertae sedis are taxa of uncertain affinities.

Sphenophyllum majus Brongniart: a small, vine-like horsetail from the Middle Pennsylvanian Pleasanton Group near Warrensburg, Missouri.

Annularia sp.: a leaf whorl of a diminutive horsetail from the Braidwood Flora of the Mazon Creek concretionary biota southwest of Chicago, Illinois.

Annularia stellata Schlotheim: a leaf whorl found at the articulation of the small horsetail found in the Cabaniss Formation at Oswego, Kansas.

Cordaites sp.: leaves at lower left from the Stranger Formation near Ottawa, Kansas. Width of leaves approximately 5 cm.

Cordaianthus sp.: a cordaitean strobilus, or cone, exhibiting fertile shoots and bracts.

Though just a tiny sprig, the presence of *Walchia* sp., an early conifer, heralds a new direction for plant life.

Samaropsis sp.: a winged seed from the Caseyville Formation in Brown County, Illinois. Taylor, et al. 2009, associates *Samaropsis* seeds with **Vojnovskya**, a gymnosperm of obscure affinities. (The fragmentary and thin lepidodendron leaves to the left of the winged seed are 4 mm in width.)

A few *Cordaicarpus* seeds, heart-shaped seeds with rounded apices along with *Cordaites* leaf fragments from the Winterset Limestone of Kansas City, Missouri.

Tim Northcutt lives in the Kansas City area and has been extensively collecting fossil plants from late Pennsylvania strata of western Missouri and eastern Kansas. He has found a number of previously unknown fossil plant bearing horizons--some of the plants from these beds are presented in this chapter. Tim is also a founding member of the Kansas Paleontology Society.

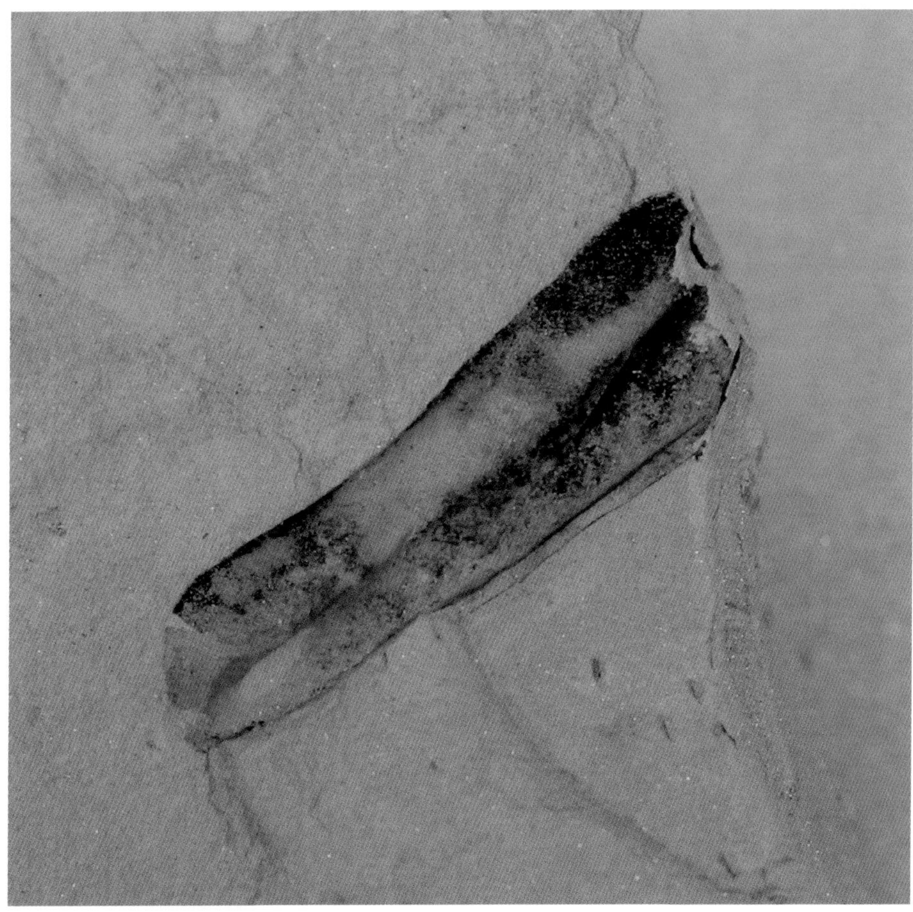

An unidentified element of the Parkville Flora thought by some to be a seed; others have conjectured possibly a cockroach egg case, or ootheca.

Index

More **Schiffer** Books *by the Author*

www.schifferbooks.com

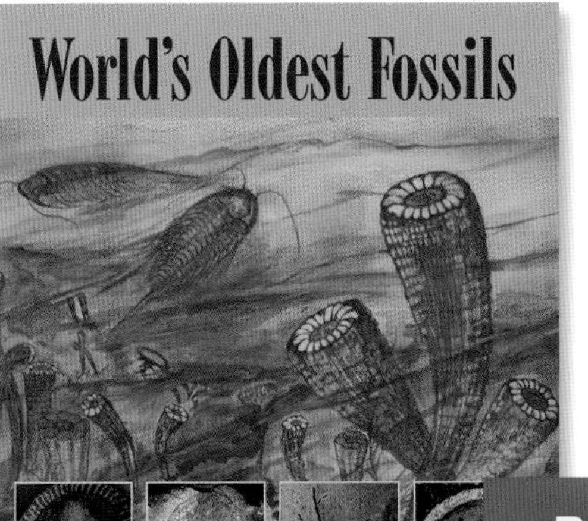

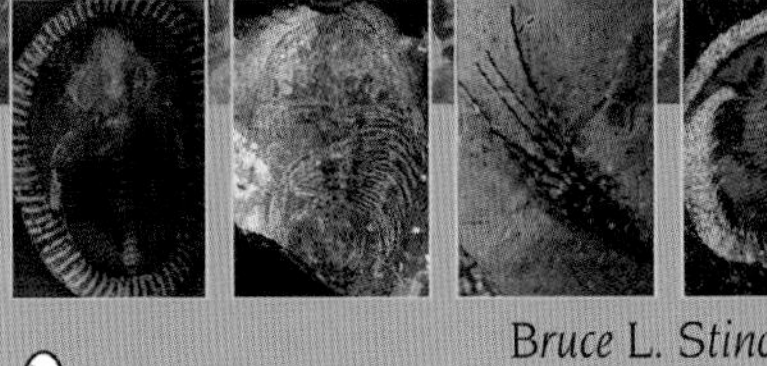

World's Oldest Fossils. Bruce L. Stinchcomb. Covers the earliest fossil record of life from strata called Precambrian and Cambrian by geologists. Illustrated and discussed are structures formed by very primitive photosynthetic life forms, Cambrian trilobites, problematic fossil-like objects, and fascinating paleontological "puzzles." Includes collector information, glossary, value ranges.

Size: 8 1/2" x 11"		509 color photos
Price Guide		160 pp.
ISBN: 978-0-7643-2697-4	soft cover	$29.95

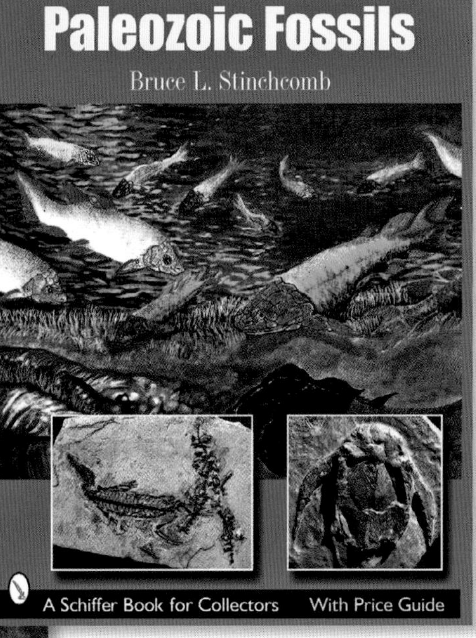

Paleozoic Fossils. Bruce L. Stinchcomb. The rich fossil record of the Paleozoic Era (545 to 300 million years ago) illustrated with 650 high quality color photos and detailed with a very readable text. The photos show the fossil remains of the earth's early animals and plants. A great book for fossil hunters of all ilks interested in early remnants of life.

Size: 8 1/2" x 11"		659 color photos
Value Guide	160 pp.	soft cover
ISBN: 978-0-7643-2917-3		$29.95

More Paleozoic Fossils. Bruce L. Stinchcomb. Over 840 Paleozoic specimens are provided, organized by biologic (taxonomic) position. Discover the sponges, archaeocyathids (reef builders), cnidaria, worms, brachiopods, bryozoans, mollusks, arthropods, echinoderms, and chordates that populated the oceans and inhabited the land from 535 to 235 million years ago. This is the period when clear evidence for plants and animals appears in hard rock.

Size: 8 1/2" x 11"	848 color photos	160 pp.
ISBN: 978-0-7643-4030-7	soft cover	$29.99

Published by Schiffer Publishing, Ltd.
4880 Lower Valley Road
Atglen, PA 19310
Phone: (610) 593-1777; Fax: (610) 593-2002
E-mail: Info@schifferbooks.com

Printed in The United States of America